Solvent Spun Rayon, Modified Cellulose Fibers and Derivatives

Solvent Spun Rayon, Modified Cellulose Fibers and Derivatives

Albin F. Turbak, EDITOR

ITT Rayonier, Inc.

A symposium sponsored by the
Cellulose, Paper, and Textile
Division at the 173rd
Meeting of the American
Chemical Society, New Orleans,
La., March 21–23, 1977

\CS SYMPOSIUM SERIES **58**

AMERICAN CHEMICAL SOCIETY
WASHINGTON, D. C. 1977

Library of Congress CIP Data

Solvent spun rayon, modified cellulose fibers and derivatives
   (ACS symposium series; 58)

   Includes bibliographical references and index.

   1. Rayon—Congresses.
   I. Turbak, Albin F., 1929-   . II. American Chemical
Society. Cellulose, Paper, and Textile Division. III.
Series: American Chemical Society. ACS symposium
series; 58.

TS1688.A1S64   677'.46   77-12220
ISBN 0-8412-0388-1   ACSMC8   58 1-269 (1977)

Copyright © 1977

American Chemical Society

PRINTED IN THE UNITED STATES OF AMERICA

ACS Symposium Series

Robert F. Gould, *Editor*

FOREWORD

The ACS SYMPOSIUM SERIES was founded in 1974 to provide
a medium for publishing symposia quickly in book form. The
format of the SERIES parallels that of the continuing ADVANCES
IN CHEMISTRY SERIES except that in order to save time the
papers are not typeset but are reproduced as they are sub-
mitted by the authors in camera-ready form. As a further
means of saving time, the papers are not edited or reviewed
except by the symposium chairman, who becomes editor of
the book. Papers published in the ACS SYMPOSIUM SERIES
are original contributions not published elsewhere in whole or
major part and include reports of research as well as reviews
since symposia may embrace both types of presentation.

CONTENTS

PREFACE

R esearch and development on new products or improved processes for converting natural based materials such as cellulose are now more important than ever, and shrewd managers are scheduling increasing efforts in the cellulose area. It is indeed ironic that the "miracle" synthetic fibers are operating at record losses while rayon and cotton are enjoying rising sales and improved market status. While this may be upsetting to those "prophets of doom" who have repeatedly forecast the demise of rayon and cellulosic textiles, it comes as no surprise to wiser, more experienced textile people who long ago recognized the many advantages which cellulosic-based materials have to offer.

This symposium, held at the New Orleans American Chemical Society Meeting by the Cellulose, Paper and Textile Division, attempts to highlight some of the current research underway in the following areas:

• Solvent Spun Rayon. This is the first time a symposium session was ever devoted solely to this topic, and it deals with diverse efforts to develop a totally recoverable and recyclable solvent spinning system to overcome viscose process deficiencies.

• Cellulose Ethers and Esters. This section covers first-release information on Cytrel synthetic tobacco as well as added technology on cellulose acetate and amic acid esters.

• Modified Cellulosics. This section includes details on the new Viloft hollow rayon fibers, lignin-modified rayon, microscopic techniques for nonwovens, drying of superabsorbent fibers, ultrasonic fiber treatment, and improved cellulose flame retardants.

Undoubtedly, a great deal of additional effort is underway in these areas which will be reported at future meetings. It is hoped that this volume will complement and enhance such research.

I would like to thank the various authors for their kind cooperation throughout this effort and also to thank their respective companies for encouraging and supporting participation by their personnel in this undertaking.

ITT Rayonier, Inc.   ALBIN F. TURBAK
Whippany, N.J. 07981
July 1977

Solvent Spun Rayon

Rayon—A Fiber with a Future

H. L. HERGERT and G. C. DAUL

ITT Rayonier, Inc., Eastern Research Div., Whippany, N.J. 07981

The title of this paper could perhaps more appropriately be called "Rayons - the Fibers With a Future" or facetiously, Rayon Strikes Back! The term rayon now represents many different fibers with a wide range of properties. Depending on the manufacturing process used, rayon can be similar to silk, wool, cotton or even paper. It can be weak and extremely water-absorbent or as strong as some of the strongest fibers made, including steel. It can be produced as continuous filament or as staple, cut in lengths of a few millimeters to several centimeters, straight or crimped, lustrous or dull, precolored, resistant to-water, -flame, or caustic soda, crosslinked or chemically modified. Of most importance, rayon, like cotton, is hydrophilic, bio-degradable and derived from the most abundant natural polymer in the world - cellulose.

Perusal of the pages of the major commercial chemical journals, such as the Chemical and Engineering News, Chemical Week and others, during the past several years might lead to the conclusion that the purely synthetic man-made fibers are the only textile fibers with a significant future. Several large oil companies with a major stake in supplying petro-chemical intermediates, have tried to foster that image by suggesting that suppliers of wood pulp, the basic raw material for viscose rayon, have insurmountable environmental problems and textile producers in the western world should focus their future on polyester. This hardly represents the facts. Rayon is currently a viable product and has an attractive future, especially if sufficient research and development is committed to reduction of chemical and energy usage in the viscose process or a new regeneration process can be perfected along the lines to be discussed in the subsequent papers.

The production of man-made fibers from cellulose can be traced back to Audemar's discovery of nitrocellulose in 1855, commercialized as filaments by deChardonnet(1) in 1889. Ralph Nader would have had a ball with this product! Because of its flammability, it had a predictable and unpleasant ending in 1940,

3

with the destruction by fire of the last producing factory in
Brazil.

The development of cuprammonium rayon by Despaissis(2)
occurred in time to step into the gap and flourished as artifi-
cial silk. Its importance as a textile fiber has gradually
diminished until today it represents a very small amount of the
textile fibers produced.

The viscose process, discovered by Cross, Bevan, and Beadle
in 1892,(3) was commercialized in the early 1900's - first as
continuous filament for textiles, for tire yarn and industrial
uses and then in the 1930's for staple fiber. Rayon production
in the United States peaked in the 1950's and again in the 1960's,
but in many cases, profits from rayon were used to diversify into
synthetic fibers. As a result, equipment was repaired but seldom
upgraded and R and D on rayon suffered. Since that time, new
growth markets have opened for rayon, such as in disposables and
nonwovens. Older plants have been shut down and the surviving
North American rayon industry has geared up to meet the textile
needs of tomorrow.

In other parts of the world, rayon production has had almost
steady growth. In Russia for example, production of cellulosic
fibers increased 30% during 1971-1975, with 60% of this growth
attributed to new plants. Even larger increases are projected for
the next five-year plan. New, more efficient rayon plants have
been built in Yugoslavia and Taiwan and other developing countries
are seriously considering the production of rayon. Since the
1930's, many versions of rayon have been developed through changes
in the basic viscose process using chemical modifiers, such as the
poly-glycols, amines, and formaldehyde. Other versions, which may
or may not require modifiers are the polynosics, which because of
high D.P. and unique structure, most closely resemble cotton in
end-use properties.

Some of the properties of the various rayons and other major
textile fibers are shown in Table I. It is seen that many of the
properties of the rayons overlap those of the other major textile
fibers. In Figure 1 this is further illustrated by the stress-
strain properties of these fibers.

At this point in our presentation, we had planned a table
showing what might be considered the properties of an "ideal"
rayon fiber. After thinking this over, the question kept popping
up - ideal for what? A list of desirable properties for a textile fiber might include:
1. Adequate tenacity (both conditioned and wet).
2. Sufficiently high wet modulus and resilience to afford
    dimensional stability in fabrics.
3. Toughness for resistance to abrasion.
4. Adequate crimp level to permit ease of processing and
    cover and loft in fabrics.
5. Hydrophilicity for comfort.
6. Resistance to sunlight, common chemicals, and laundering.

TABLE I

PHYSICAL PROPERTIES OF MAJOR TEXTILE FIBERS

	Polynosic Rayon	Cotton	HWM Rayon	Polyester	Nylon-6	Regular Rayon
Tenacity						
Cond. g/d	3.5-5.0	1.8-3.2	3.0-4.5	2.4-7.0	3.5-7.2	0.7-3.2
km	32-45	16-29	27-41	22-63	32-65	6-29
Wet, g/d	2.5-4.0	1.6-3.2	2.0-3.5	2.4-7.0	2.5-6.1	0.7-1.8
km	23-36	14-29	18-32	22-63	23-55	6-16
Elongation, %						
Cond.	8-10	7-9	10-12	12-55	30-90	15-30
Wet	10-14	8-10	12-15	12-55	42-100	20-40
Wet Tenacity at 5% Extension, g/d	0.5-2.0	0.5-1.0	0.5-0.7	0.3-0.5	0.1-0.5	0.1-0.3
km	4.5-18	4.5-9.0	4.5-6.3	2.7-4.5	0.9-4.5	0.9-2.7
Moisture Regain, %	10-14	7-9	10-14	0.4	3-5	10-14
Water Retention, %	60-70	45-55	60-80	-	4-10	90-110

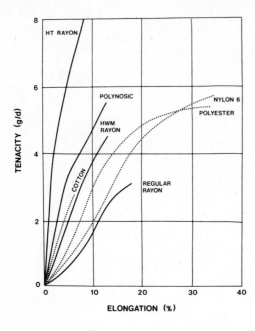

Figure 1. *Fiber stress-strain curves (conditioned)*

TABLE II

WORLD FIBER CONSUMPTION[*]

	1972-4	1985	Change (%)
Population (mill	3817	4759	24.6
Per Capital Fiber Consumption (Kg)	6.9	8.4	21.7
Fiber Consumption (000 tons)	26,978	40,230	49.1
Man-made			
Rayon and acetate	3,564	4,190	17.6
Non-cellulosic	7,676	18,015	34.7
Natural			
Cotton	12,970	15,695	21.0
Wool	1,985	1,632	5.1
Other	711	698	-1.8

* American Dyestuff Reporter, June, 1976

7. Dyeability.
8. Flame retardancy.
9. Uniformity and cleanliness of product.
10. Price stability.

Obviously, there can be no one fiber with all these properties and for the majority of end-uses, not all are needed. Therein lies the strength of the large family of rayon fibers which can be practically tailor-made or engineered to suit specific end-uses. The textile industry of today is, of necessity accepting this fact and is using combinations of fibers in blends chosen to produce the fabrics which are most aesthetically appealing, useful, serviceable and profitable. For apparel, blends of cellulosics and synthetics to provide both comfort and utility are the accepted norm. We believe the use of 50% or more cellulosic fiber with polyester is a practical necessity to prevent discomfort, soiled clothes, and the embarrassment of wet spots on plastic seat covers.

Building water absorbency into polyester has been a longtime goal of this fiber industry but has usually been at the sacrifice of some of the important, desirable attributes of this fiber, and results in increased cost of production. We believe that the most intelligent and economic approach to the textiles of the future is to blend the synthetics with cellulose, to get the appropriate balance of strength, easy care, and moisture absorption.

In the June, 1976, American Dyestuff Reporter, predictions of world fiber consumption relative to world population growth through 1985 were given. (Table II). These figures show an increase of 17.6% in the production of rayon and acetate. Rayon, of course, will represent the larger proportion of this growth. We believe these figures are conservative for man-made cellulosics relative to synthetics and that the divisions could be more favorable to rayon and acetate provided cost-effective improvements are made to the viscose process or an alternate low-capital approach to forming regenerated cellulosic fibers is found.

The growth in world population will obviously result in a demand for food production at the expense of cotton, especially in countries such as India and Pakistan, where there is a burgeoning population with increased textile and food needs.

Rayon - Its Problems

Considering rayon as a fiber with a future, it is necessary to look at the current problems associated with its manufacture. Simply put, in today's economy and with emphasis on energy conservation and environmental protection, the main problems related to the production of fibers by the viscose process are:
1. Undesirable air and water emissions, BOD, H_2S, zinc.

2. Intensive capital requirement.
3. Large energy requirements.
4. Relatively large labor requirements in most existing plants.
5. Increasing cost of raw materials.

Other problems related to the effects of meeting environmental protection regulations can have adverse effects on quality and availability of raw materials.

Rayon - Its Opportunities

The above problems seem insurmountable, but problems are usually followed by opportunities, and such is the case here. Major advances are being made in modification and redesign of viscose rayon plants to reduce pollutants, for example, by recovery of zinc by ion-exchange, crystallization, or other techniques; use of more purified celluloses; reduction of gaseous emissions by absorption or scrubbing; more efficient fiber washing and drying techniques, and the like. Large scale plants can be designed to produce fibers for more specific, major end-uses and with fewer product lines which will be more efficient in use of labor and capital.

The major advantages possessed by regenerated cellulose fiber production of today are:
1. Availability of major raw materials.
   a) Cellulose - nature's renewable polymer.
   b) CS_2 - recoverable to the extent of 50%.
   c) Sulfuric acid and caustic soda normally in large supply.
2. Non-dependence on oil.
3. Price stability relative to cotton.

These advantages have been recognized in other countries with the USSR as a prime example. In the May issue of Khimischeskie Volokna, Shimko(4) describes the technical progress in the cellulosic fiber industry and projections for the USSR's 10th five-year plan. He states: "The production of cellulosic fibers is planned to increase further in the 10th five-year plan, the main reason being: the useful properties, especially the physiological ones, of those fibers and the mass-market articles produced from them. An increase in the output of cellulosic fibers will not only help to overcome the shortage of hygroscopic fibers but will also result in a substantial savings of material and labor resources. The fact that the USSR possesses a huge self-renewing source of the starting material (i.e., wood) is a further factor favoring an increase in the production of cellulosic fibers."

He further states, "the principal directions of technical progress in the production of viscose rayon staple fibers are: the production of high-modulus and polynosic fibers and higher output

of better quality viscose rayon staple - high-modulus and polynosic fibers - like cotton, possess good strength and a high
modulus when wet. Their wet strength is better than that of
(regular) rayon staple. Although the cost-effectiveness of the
production of high-modulus fibers is not very favorable, their
use in the national economy, in place of cotton, is advantageous."

In Russia, as in other countries, aside from economics and
raw material availability, there is a growing awareness that the
comfort (physiological) factors inherent in cellulosic fibers are
essential in blends with the synthetics to overcome this major
deficiency in these fibers.

Even assuming opportunities for further expansion of viscose
rayon production in the USSR and elsewhere, and for improving the
economics of the viscose process, we, however, believe that significant future expansion for rayon, or man-made cellulosic
fibers, will result with the development of a totally new process,
designed to overcome the deficiencies mentioned before.

One approach to a new process involves the use of a recoverable and recyclable solvent (or combination of solvents) which
will dissolve cellulose by a simple mixing step followed by extrusion of the solution to form a fiber. A closed-loop system,
which would emit practically nothing to the atmosphere or water,
would eliminate pollution problems. Several leads in this direction have been developed at our laboratories in Whippany, New
Jersey, and will be described in detail in subsequent papers on
this program. However, much more needs to be done to reach the
ultimate goals.

Ideally, such a process should involve use of simple cellulose solutions of higher concentrations than those presently used
(6-9%), use of low boiling solvents to minimize energy required
for recovery, spinning at high speeds similar to those used in
the acetate and synthetic fiber industries, and purification without the use of excessive amounts of water for washing or large
amounts of heat for drying. Such a plant would, therefore, have
most of those attributes required to overcome the present-day
problems associated with the rayon industry. This concept should
present a real challenge to cellulose research for the coming
years and the rewards would be enormous. Some of the requirements for an ideal cellulose (rayon) fiber production process are
shown in Table III.

Other approaches are, of course, possible. One such is that
envisioned by the eminent polymer scientist, Dr. Herman F. Mark,
who recently said(5), "For the future, continued research efforts
will be greatly influenced by the fact that cellulose is the organic substance produced in greatest quantity by nature - a renewable resource that contrasts sharply with coal, oil and gas,
the reserves of which are being seriously diminished year-by-year.
It would indeed be a wonderful success story for human endeavor,
if, in the future, we can combine our present knowledge of the
structure and reactivity of cellulose with a better understanding

TABLE III

REQUIREMENTS FOR IDEAL RAYON FIBER PRODUCTION PROCESS

A. Renewable/recoverable raw materials.

B. Low Capital and labor costs, low energy requirement.

C. Minimal effect on environment.

D. Simple solution-making process.

E. Dry and/or solvent spinning.

F. High conversion rate of spin-dope to fiber (includes
high cellulose/solvent ratio).

G. Ability to produce a uniform product that will require minimal handling from fiber to end product.

of its biosynthesis, in order to produce the cellulose molecule directly from water and CO_2, with the aid of sunlight, at a rate 5 to 10 times greater than that which occurs in nature."

Controlled growth of cellulose in hydroponic factories to produce fibers directly, could be the fulfillment of such a wish or alternately, a means to further extend the capabilities of the world's forest resources to furnish cellulose to produce rayon for the multitude of end-uses required by mankind. Whether it will be made by the viscose process, or spun from solvents, or by some yet to-be-discovered process, or by all three, remains to be seen. To paraphrase a popular automobile commercial --

"There Will Be A Rayon In <u>Your</u> Future"

Abstract

Modern textile blends continue to require the aesthetics and moisture-absorbing properties provided by cellulose in the native state (cotton) or in regenerated form (rayon). Escalating prices and problematical availability of some intermediates for purely synthetic textiles, coupled with gradual conversion of cotton-growing land to food production, suggest a careful re-examination of the future of rayon. Wood cellulose, caustic soda and carbon disulfide, the major raw materials for rayon production by the existing viscose process, are not dependent upon oil and will continue to be available in ample supply. On the other hand, the viscose process is energy intensive and has emission problems. Significant expansion of the rayon industry will, therefore, require development of a totally new process. The needs of such a process in terms of raw materials, type of spinning, investment costs and fiber properties will be detailed in this paper.

Literature Cited

1. Chardonnet, H., French Pat. 165,349 (May 12, 1884).
2. Despaissis, L. H., French Pat. (1890).
3. Cross, et al, British Pat. 8700 (April 8, 1893).
4. I. G. Shimko Khimicheski Volokna, No. 3, pp 8-12, May-June 1976.
5. Mark, H. F., Celluloses - Past Present and Future, 50th Anniversary Lecture, Nov. 26, 1975, Dorval, Que. Can.

2

A Critical Review of Cellulose Solvent Systems

A. F. TURBAK, R. B. HAMMER, R. E. DAVIES, and N. A. PORTNOY

ITT Rayonier, Inc., Eastern Research Div., Whippany, N.J. 07981

Cellulose is the most abundant renewable, organic raw material available in the world today. Yet, for all it's availability, it has still not reached its potential utility in many areas of application. One of the major reasons for this is that many end-use applications require that cellulose be in a different form from that found in nature. In most of these applications, it is necessary first to dissolve cellulose in some manner and then to re-form it from such solutions into the desired products. It is this very important dissolving step which has proved to be either cumbersome or expensive compared to alternate materials which compete for market positions. In many cases only cellulose has the desirable properties required for end product use and, in these instances, the methods required to achieve cellulose solution present potential hazards and pollution control problems. Thus, improved techniques for dissolving cellulose are urgently needed if cellulose is to continue to occupy a competitive market position. This is particularly true of the viscose industry where pollution control continues to place ever increasing restraints on the process.

Scientists have long recognized the need for more efficient cellulose solvent systems and hundreds of publications have been issued covering a wide range of approaches. Several excellent review articles on swelling and dissolving cellulose by Warwicker (1,2) Jayme (3), Phillip (4,5), Polyola (6) and Brandrup (7) have been published and we shall not attempt simply to resubmit their data. Rather, this paper shall attempt to consider the various processes from the commercial as well as the scientific viewpoint to emphasize potential areas for contribution which still exist, particularly for systems capable of producing cellulosic fibers through the use of organic solvents.

Fundamentally, all of the known methods for dissolving cellulose can be summarized under four main categories:

A. Cellulose As A Base
B. Cellulose As An Acid
C. Cellulose Complexes
D. Cellulose Derivatives

While each of the above areas has received extensive study and
some have been commercialized, several of the approaches are
really not completely understood. Still others involve mixtures
of many ingredients and could never be valuable industrially.

In trying to explain why certain materials do in fact dissolve cellulose, several authors have simplistically indicated
that what is required is that the solubility parameter of the
solvent must be the same as that of cellulose. While such state-
ments are directionally correct, they are incomplete and, perhaps, even a little misleading, since many solvents having the
proper solubility parameter values will not dissolve cellulose.
For example, DMF and DMSO both have solubility parameters in the
range calculated for cellulose but neither of these dissolve
cellulose under any known conditions.

The solubility parameter (δ) or cohesive energy density
(CED) values relate directly to the energy of vaporization
which is easily obtainable for many liquids, and will apply
directly to the miscibility of liquids or amorphous polymers
where (δ) values will relate to simple heats of mixing.

When dealing with polymers in general, other factors may be
equally as important, or even more important than single solubility parameter values. For example, with crystalline polymers,
the heat of fusion or melting energy, must also be considered as
an entirely separate factor. Furthermore, superimposed on these
factors is the hydrogen bonding capability of the polymer in
question which may not be consequential for polymers such as
crystalline polypropylene, but which is extremely important with
natural polymers such as proteins and cellulose in particular.
These factors, as well as second order effects were recognized
many years ago by Spurlin, Burrell, and Barton and several excellent reviews are available for further reference (8,9,10).
A combination of such factors must be considered in trying to
find a solvent for cellulose and help to give guidance as to
conditions as well as compounds which must be employed if usable
cellulose solutions are to be obtained.

Even if a material is a satisfactory solvent for cellulose,
it is necessary to emphasize two other important factors which
must be seriously evaluated for any potentially commercial cellulose solvent system, these are "recovery" and "recycle". The
recovery and recycle factors, perhaps more than any others, are
the most important ones which, in the final analysis, decide
whether or not any particular approach will be economically
feasible.

One final factor must also be kept in mind relative to
evaluating all of the data that appear in the literature re-
garding cellulose solvents. Specifically it must be emphasized
that vast differences exist between being able to coagulate or

precipitate a solid mass, as compared to truly spinning fibers
with reasonable physical properties.

Keeping these factors in mind, let us examine the four categories listed above which encompass essentially all of the cellulose solvent systems reported in the literature.

A. Cellulose As A Base

All alcohols, whether polymeric or not, can be forced to act
as acids or bases relative to other reagents, depending upon the
strength of the reagent employed. Thus, cellulose will act as a
base and become protonated by protonic acids or will act as a
"Lewis" base and supply electrons from its oxygens to electron-
receptive centers on "Lewis" acids.

1. Protonic Acids (phosphoric, sulfuric, nitric). The
ability of 78% phosphoric acid, 68% nitric acid, 42% hydrochloric
acid or 70% sulfuric acid to dissolve cellulose rapidly at room
temperature is well known. While the literature is full of reports in which the resulting solutions are referred to as "hy-
drates", they can also be regarded as protonated hydroxyls where
the specific reagent concentrations employed were the ones needed
to provide sufficient acid strength to protonate the cellulose so
that the resulting positively changed $ROH_2^{\oplus}$ cellulose moiety is
able to dissolve in the excess reagent. In these cases, recovery
and recycle of the concentrated acids is the limiting factor,
economically. For example, excellent cellulose solutions can be
rapidly prepared at room temperature using phosphoric acid.
However, no one has yet suggested a feasible approach to casting
the cellulose and recovering the phosphoric acid without neutralization. In our laboratory, we have tried to precipitate such
phosphoric acid solutions into glacial acetic acid in the anti-
cipation that the acetic acid could be volatilized, recovered
and recycled, along with the released pure phosphoric acid. The
initial results were not too successful since the precipitation
was too slow to be useful under the particular conditions employed. However, the concept of using a weaker, volatile acidic
material to coagulate strong acid cellulose solutions does represent a novel approach to developing a total recyclable cellulose solvent system.

2. Lewis Acids (zinc chloride, thiocyanates, iodides, bro-
mides). The ability of certain salt solutions to swell and
dissolve cellulose is also reported (11). At the high concentrations normally necessary to achieve solution, these salts not only
provide necessary acid functionality but also significantly alter
the ionic nature of the aqueous mediums so as to further assist
dissolution. The need for having acid salts is demonstrated in
the case of thiocyanates where the sodium, potassium and ammonium
thiocyanates do not dissolve cellulose while the calcium and

strontium salts do give cellulose solutions up to about the 400 D.P. level indicating that it is the thiocyanic acid formed in solution which is actually involved in the dissolving action.

B. Cellulose As An Acid

Perhaps more work has been undertaken on swelling cellulose with bases than with any other class of chemical compounds. The hydrogen atoms on the hydroxyls of cellulose, like most alcohols, have a degree of acid character and therefore readily interact with reasonably strong inorganic and organic bases.

1. Inorganic Bases (sodium zincate, hydrazine, inorganic hydroxides).
It should be immediately noted that the concentration of base needed for obtaining maximum swelling of cellulose by osmotic action is not the same concentration that is required to effect actual compound formation. Thus while 8-10% NaOH exhibits the maximum swelling of cellulose fibers, concentrations of about 18% NaOH are actually required to form sodium cellulosate structures. Other inorganic bases like potassium and lithium hydroxide have also been used similarly for their swelling action. Various additives to the caustic soda, such as zinc oxide, have been used to make sodium zincate solutions which enhance the action of caustic so higher D.P. materials can be dissolved. The temperatures of these aqueous bases play a further dominant role in the swelling and dissolving phenomenon; with the colder conditions giving more swelling and dissolution. While a considerable amount of effort has been devoted to aqueous alkali systems, they normally do not display sufficient dissolving power to completely overcome the crystal and hydrogen bonding energies of cellulose to give acceptable solutions of higher D.P. materials of interest for direct commercial conversion. In addition, these aqueous alkaline systems present several recovery and recycle problems.

The use of the inorganic base, hydrazine, for swelling cellulose was considered by Hess and Trogus many years ago.(12) However, they were never able to obtain cellulose solutions under the conditions they employed. More recently, Litt (13) has reported conditions under which he has been able to obtain complete solutions of high D.P. cellulose in hot (150°-200°C) hydrazine and 207-345 kPa (30-50 psi) pressure. These results serve to demonstrate that several factors are important for a swelling reagent to become a solvent. The original investigations in 1931 by Hess employed hydrazine hydrate but did not employ the more extreme conditions of heat and temperature recently employed by Litt in 1976 to obtain solutions. Litt was able to get solutions of cellulose using either pure hydrazine or hydrazine hydrate providing he employed elevated temperatures and pressures. It was noted previously that crystalline forces

and hydrogen bonding could be important factors and this case
clearly demonstrates that even a material having the proper solubility parameter had to be employed under forcing conditions to
overcome these factors to achieve solution. It has yet to be
demonstrated, however, if textile quality yarns or packaging
quality films can be prepared from such hydrazine solutions.
Certainly, to our knowledge no data along these lines have yet
been published.

   2. Organic Bases (Triton-quaternary hydroxides, amines,
DMSO/CH$_3$NH$_2$, amine oxides). The ability of cellulose to act
as an acid towards various organic bases has also been evaluated.
Perhaps the best known organic bases are the various "Triton"
bases which are quaternary ammonium hydroxides and the best solvent of this series is "Triton B" or benzyltrimethylammonium hydroxide. It is interesting again to note here that the benzyl
grouping is far more effective than other groups such as methyl,
ethyl or propyl and this improvement has been attributed to the
fact that the bulkier benzyl group behaves like a "wedge" to
effectively separate the cellulose chains once it has entered the
crystalline region.

   A considerable amount of work has also been reported by
Segal and others (14) relative to the use of various amines and
diamines for swelling cellulose. None of these systems were
claimed to cause sufficient swelling to give cellulose solutions.
More recently, Phillip and his co-workers (15,16) have reported
that 16.5% methylamine in DMSO gave better cellulose solutions
than any other amine or any other concentration of methylamine
used. The cellulose is claimed to dissolve by reaction with an
equimolar complex of CH$_3$NH$_2$/DMSO and $\sim$80% of the introduced
cellulose dissolved under cold anhydrous conditions. As ex-
citing as this seems at first glance, dissolution of only 80% of
the cellulose may not be of value for commercial consideration.
Unless the dissolution of cellulose is actually better than 99%,
the system would certainly be too expensive to use relative to
yield and filtration costs. The specificity of CH$_3$NH$_2$ and in
particular the 16.5% concentration is intriguing and deserves
more study. Perhaps, this system might more properly belong
under consideration as an organic complex rather than as a pure
base system, but more data are needed to firmly decide the exact
nature of the reaction.

   One other organic base system which dissolves cellulose as
an acid involves compounds not normally considered as bases, but
which in fact, are very good "Lewis" bases. In 1939, Graenacher
and Sallman (17) reported that aliphatic and cycloaliphatic amine
oxides such as triethylamine oxide or cyclohexyl dimethyl amine
oxide gave 7-10% solutions of cellulose at 50-90°C. More recently, Johnson (18) has reported that alicyclic amine oxides
such as N-methylmorpholine oxide give up to 6% solutions of cellulose at 110°C. These materials are most interesting and could

offer commercial possibilities if adequate recovery and recycle
procedures could be established. To our knowledge, no one has
yet published any actual fiber or film data on products derived
from such systems.

Thus, a wide range of inorganic and organic bases have been
examined for dissolving and swelling cellulose. Several systems
have achieved a sufficiently good balance of properties to provide good cellulose solutions and it is expected that further
efforts will be forthcoming to achieve acceptable recovery/re-
cycle and physical properties which could lead to industrial
acceptance.

C. Cellulose Complexes

1. Inorganic Complexes (cuene, cadoxene, EWNN, cuprammonium). The use of various copper complexes to dissolve cellulose is well known. More recently, complexes of cadmium and iron
have been added to the list to reduce the sensitivity of such
cellulose solutions to degradation by exposure to air. This work
is reviewed rather completely by Jayme (3) who, with his coworkers, has contributed extensively to this area.

Inorganic metallic complexes such as cuprammonium have met
with only minimal commercial success for production fibers and
films for two reasons. First, complete recovery of metallic effluent contaminants is difficult at the extremely low ppm levels
needed to meet pollution requirements and secondly the overall
economics and fiber properties (of the cuprammonium process, at
least) were not as good as those of the viscose process. The
second factor would today be less important than the recovery and
pollution aspect and, until this aspect is solved, metal inorganic complexes will continue to have limited utility for film
and fiber production.

2. Organic Complexes (DMSO/CH_3NH_2, (HOCH$_2$CHOHCH$_2$S)$_2$. The
possible use of organic complexes to dissolve cellulose appears
to be extremely limited. Of all the systems reported, only two
seem to qualify as organic complexes. The first of these uses
DMSO/16.5% CH_3NH_2 and was previously discussed under category B
(above) since its action may be related more to its basic nature
than to a true complex formation.

The second system was reported by Petrov in 1965 (19) who
claims that cellulose dissolves directly at 110°C in a neutral
solvent - bis (β-γ dihydroxypropyl) disulfide - prepared by
oxidation of thioglycerol. This solvent works well for dissolving cellophane films at 300-600 D.P., but is not good for
directly dissolving higher D.P. cotton linters or regular purified wood pulps. In any case, it is intriguing to find a neutral
solvent that appears to be capable of directly dissolving reasonably high D.P. cellulose and further work should be undertaken to
ascertain why this particular structure seems to be effective.

D. Cellulose Derivatives

1. Stable Derivatives (esters, ethers). A wide variety of
stable cellulose derivatives are known and used commercially.
The esters and ethers have found scores of commercial uses but
only the acetate and nitrate have ever been utilized for producing regenerated cellulose products; the acetate to give Fortisan
and the nitrate to make regenerated tubular cellulose films.

2. Unstable Derivatives.

    a. Sulfur   (xanthates, SO_2/amines - sulfites)

    b. Nitrogen (DMF/N_2O_4, $DMSO/N_2O_4$ - nitrites)

    c. Carbon   (carbonates, formates, $DMSO/(CH_2O)x$ -
                methylol)

Unstable cellulose derivatives have been and are being
actively investigated in depth for use in preparing regenerated
fibers and films. This paper will consider such "transient"
derivatives under three main headings: a) sulfur- b) nitrogen-
and c) carbon-containing intermediates.

a) Sulfur Intermediates: The use of CS_2 to form xanthates
some 90 years ago still forms the backbone of the present day
rayon industry. Improved processes are urgently needed which
can not only overcome the various pollution problems associated
with the viscose process, but which might also hopefully lower
both the initial capital investment and subsequent operating
costs associated with present day rayon plants.

One attempt along these lines is reported by Kimura et al
(20) who used a modified organic type system for xanthation.
They used $DMSO/CS_2$/amine to dissolve cellulose and reported
fibers having conditioned and wet tenacity/elongation of 3.0
g/d/14% and 1.8 g/d/25% respectively. Their coagulation and regeneration steps did not involve the use of acids and thus would
significantly reduce part of the pollution load normally associated with rayon production. However, initial laboratory attempts
to reproduce the reported process gave rise to an extremely nox-
ious odor and this could represent sufficient detriment to offset
other possible advantages.

As another possible approach to dissolving cellulose, several investigators have studied the use of SO_2/amine solvent systems. Extensive work in this area is reported by Phillip and
his coworkers (21) by Yanazaki and Nakao (22) and by Hata and
Yokota (23-29). Their efforts covered a wide range of amines and
organic solvent diluents which were both polar and non-polar.

While initially it was thought that some type of complex was
being formed with the cellulose, it is now reasonably established that this treatment results in the formation of amine
salts of the rather unstable cellulose sulfite ester. Again
filaments reportedly were spun, but no fiber physical properties
were reported. The use of sulfite esters for preparing rayon in
preference to the viscose system would critically depend on good
recovery of all starting materials, and this may be an area
worthy of both chemical and engineering effort.

   b) Nitrogenous Intermediates: The use of unstable nitrogenous intermediates for achieving cellulose solution in organic
solvents is concentrated mainly in the area of systems involving
nitrogen oxides. Initially nitrogen oxides were used to oxidize
cellulose to the 6-carboxy derivative. The original work of
Kenyon, Yackel, Unruh, Fowler, and McGee in the 40's stands as
a milestone in this area. (30–34) More recently, Pavlyuchenko
and Ermolenko and coworkers have presented further detailed
studies on the use of N_2O_4 for cellulose oxidation. (35–39) In
1947, Fowler et al (34) reported that various solvents, when used
in conjunction with N_2O_4, gave different responses to the action
of this reagent on cellulose and found that many solvents actually gave rise to solutions of cellulose without high degrees of
oxidation occurring. They used over 40 different solvents where
the N_2O_4 to solvent ratios were at least 1/1 or higher and
finally concluded that the amount of oxidation vs. cellulose
solution was related to the polarity of the solvent employed.
With non-polar solvents such as CCl_4 etc. the N_2O_4 produced
mostly oxidation while the more polar solvents gave rise to cellulose solutions with greatly diminished oxidation. If no solvents were employed and large excesses of cold liquid N_2O_4/N_2O_3
mixtures were used, then cellulose dissolved with very little
oxidation and could be recovered from such solutions essentially
chemically unchanged except for a D.P. loss. This was first reported by Hiatt and Crane in 1949 (40) and subsequently studied
in great detail by Chu in 1970. (41)
   Following these classic disclosures by Fowler, other re-
searchers began to examine even more polar solvents for use with
N_2O_4. Thus, Williams (42) studied the use of $DMSO/N_2O_4$ as a
solvent system for cellulose. It was later demonstrated rather
clearly by Hergert and Zopolis (43) that the $DMSO/N_2O_4$ system
would dissolve cellulose more effectively if there was a small
amount of water present to keep the cellulose structure open for
reaction. Surprisingly very little more work has been done with
the $DMSO/N_2O_4$ system up to the present time.
   Subsequently, Nakao obtained patent coverage on the use of
DMF/N_2O_4 mixtures to dissolve cellulose (44,45) and on the addi-
tions of a wide range of other polymers in DMF to such cellulose
solutions to produce special types of products. Mahomed (46)

describes the use of cellulose/DMF/N_2O_4 solutions as coating materials for glass fibers to improve product performance.

During this period several other workers began to appreciate that very polar solvents gave cellulose solutions from which cellulose could be recovered in an essentially unchanged state rather than in the highly oxidized state previously associated with N_2O_4 treatments. A considerable amount of research was published trying to define what was occurring when the N_2O_4 was used with the polar solvents. Clermont and Bender and their associates reported studies where cellulose was dissolved in both cold and hot solutions with N_2O_4/DMF mixtures. (47-49) They found that essentially unchanged cellulose with no added nitrogen or carboxyl levels was obtained from cold dissolution, while water-soluble cellulose resulted from increased dissolving temperatures apparently due to the formation of various D.S. cellulose nitrates. They subsequently tried to use this procedure to dissolve lignocellulose removed from wood chips.(50)

In a series of articles dating from 1969-1976, Schweiger has been reporting studies aimed at trying to elucidate what was happening in the DMF/N_2O_4 treatment of cellulose and tried to use such solutions for forming other derivatives. (51-56) He was able to isolate a product from a pyridine-modified dope which appeared to be an unstable cellulose nitrite intermediate since it could produce alkyl nitrites when decomposed by lower molecular weight alcohols. While this is not a direct structural proof it certainly is sufficient to substantiate his proposal that cellulose reacts with N_2O_4 to give cellulose nitrite and HNO_3 rather than forming some type of N_2O_4/cellulose association complex. The cellulose nitrite is subsequently rapidly decomposed by protonic solvents to regenerate the cellulose and give HNO_2 along with HNO_3 for recovery and recycle. The structure and reactions of N_2O_4 in general have been reviewed by Gray (52) while the reactions of N_2O_4 to nitrate and nitrosate alcohols and amines are reported by White and Feldman.(58)

Pasteka and Mislovicova studied the effects of various dissolving conditions on the D.P. loss of cellulose in the DMF/N_2O_4 system. (59-61) They noted that the moisture content of the system and even the rate of stirring caused a drop in D.P. and that the presence of pyridine or $(C_2H_5)_3N$ did not inhibit such loss. While moisture may well relate to D.P. loss, it is difficult to understand how stirring rate could have such an effect unless it was reflecting local temperature rises that occurred under the conditions employed for stirring. In any case, these investigators again confirmed no increase in either nitrogen or carboxyl content for the regenerated cellulose.

The N_2O_4 system has also been investigated by several Russians (62-64) who actually report physical properties for fibers spun from DMF/N_2O_4 and EtOAc/N_2O_4 systems. These fibers are about 120 denier and have tenacities of 1.6 g/d with 5-6%

elongations. Obviously more definition of the N_2O_4/organic solvent system is needed to determine it's potential as a solvent
based rayon process to substitute for viscose and this work will
be reported later in this symposium.

c. Carbon Intermediates: The use of non-hetero atom-con-
taining moieties to make cellulose intermediates of transient
stability has received relatively little attention in spite of
the fact that such derivatives might ultimately offer the best
prospects for non-polluting systems. Cellulose carbonate has
been prepared and reported. However, it evidently is so unstable as to have essentially no utility as an intermediate.
Also, as prepared from phosgene, it would face other industrial
problems.

Cellulose formate is well known and has been reported as
being rapidly prepared by the action of 95% formic acid on cellulose. The cleavage of cellulose formate by hot steam also represents an interesting approach for "dry spinning" this derivative. While formic acid would not be considered a "material of
choice" under most circumstances for industrial use, it is certainly no worse than "hydrazine" in most safety and health con-
siderations and this approach, or one similar to it, will undoubtedly see further effort in the future.

Recently, Nicholson and D.C. Johnson reported on their work
on dissolving cellulose in mixtures of DMSO with paraformaldehyde. (65) Initial indications are that a cellulose methylol compound is formed which is stable under the elevated temperatures
of solution preparation and is subsequently stable for days under
storage in open air at room temperature. The extreme specificity
of this combination of reagents is particularly notable. For example, Seymour and E.L. Johnson (66,67) noted that neither DMF,
DMAc, acetone, HMPA, nitromethane, acrylonitrile, acetonitrile,
nor sulfolane can be substituted for the DMSO. Thus DMSO and
only DMSO has been found to be effective to date for achieving
cellulose solutions with formaldehyde. This extreme specificity
may well relate in some way to the fact that DMSO itself breaks
down into DMS and paraformaldehyde on heating (68) or may possibly be interacting to stabilize the proposed cellulose methylol
intermediate whereas no other reagent is evidently able to do so.
It should be further noted that while only one mole of paraformaldehyde is required to hold one mole of the cellulose in solution, large molar excesses of 5/1 $(CH_2O)x$/cellulose must be employed initially for dissolution to occur. Thus this system
which appears to initially offer an easy route to cellulose
solutions, may offer considerable difficulty commercially from
the aspects of spinning, recovery and recycle. Further data
from this work are to be presented later in this symposium.

Throughout this review, an attempt has been made to highlight the need for a low investment cost, non-polluting solvent system for cellulose to substitute for the viscose process. None of the systems known to date meet the necessary requirements for commercial exploitation. However, as has always been the case, whenever a need like this exists some outstanding scientists will develop methods to produce the desired results – and this case will be no exception. The cellulose chemists are equal to the challenge and the rewards for success will be large.

Literature Cited

1. Warwicker, J.O., Jeffries, R., Colbran, R.L., and Robinson, R.N., Shirley Inst., Pamphlet No. 93 (1966), Manchester, England.
2. Warwicker, J.O., High Polymers, Vol. V, Part IV, Wiley-Interscience, N.Y., 1971.
3. Jayme, G., High Polymers, Vol. V, Part IV, Wiley-Interscience, N.Y., 1971.
4. Phillip, B., Schleicher, H., and Wagenknecht, W., Cellulose Chem. Tech., 9, 265–82 (1975).
5. Phillip, B. Schleicher, H., and Wagenknecht, W., C.A. 83 165567s; Chem. Vlakno 25 (10–22) 1975.
6. Polyola, L., and Aarnikowa, P.L., Kem.-Kemi 2, (1) 27–9 (1975).
7. Brandrup, J. and Immergut, E.H., Polymer Handbook, 2nd ed. V-101, John Wiley & Sons, N.Y. 1975.
8. Spurlin, H.M., High Polymers Vol. V, Part III, Wiley-Interscience, N.Y. (1955).
9. Burrell, H., Official Digest (726–758) 1955.
10. Barton, A.F., Chem. Reviews, 75, No. 6, (731–753) 1975.
11. Williams, H.E., J. Soc. Chem. Ind., 40, 221T, 1921.
12. Hess, K. and Trogus, C., Z. Phys. Chem., B14, 387 (1931).
13. Litt, M., Cell. Div. Preprints, ACS Mtg., N.Y. 1976.
14. Segal, L., High Polymers, Vol. V, Part IV, Wiley-Interscience, N.Y. 1971.
15. Phillip, B., and Schleicher, H., C.A., 74, 127361b (1971).
16. Koura, A., Schleicher, H., and Phillip, B., Faserforsch. Textiltech. 23, (3) 128–33 1972; C.A., 77, 50396u 1972.
17. Graenacher, C., and Sallmann, U.S. 2,179,181 (1939).
18. Johnson, D.L., U.S. 3,508,941 (1970); B.P. 1,144,048 (1969).
19. Petrov, V.G., C.A. 63, 10161g, 1965.
20. Kimura, T., Yamamura, T., Kawai, A., and Nagai, S., Japan Patent 69 02,592.
21. Phillip, B., Schleicher, H., and Laskowski, I., Faserforsch Textiltech., 23, 60–65, (1972).
22. Yamazaki, S., and Nakao, O., C.A., 81, 154860q (1974).
23. Kata, K., and Yokota, K., C.A. 66, 47464g (1967).

24. Hata, K. and Yokota, K., C.A. 70, 69191a, (1969).
25. Hata, K. and Yokota, K., C.A. 70, 69192b, (1969).
26. Hata, K. and Yokota, K., U.S. 3,424,702 (1969).
27. Hata, K. and Yokota, K., C.A., 72, 33464u (1970).
28. Hata, K. and Yokota, K., C.A., 74, 14323x (1971).
29. Hata, K. and Yokota, K., C.A., 75, 153094g (1971).
30. Kenyon, W. and Yackel, E., U.S. 2,448,892.
31. Yackel, E.C. and Kenyon, W., J.A.C.S. 64, 121 (1942).
32. Unruh, C.C. and Kenyon, W., J.A.C.S. 64, 127 (1942).
33. McGee, P., Fowler, W.F., et al, J.A.C.S. 69, 355 (1947).
34. Fowler, W., Unruh, C., McGee, P. and Kenyon, W., J.A.C.S. 69, 1636 (1947).
35. Pavlyuchenko, M. and Ermolenko, I., C.A. 1739d (1956) Izvest. Akad. Nauk. SSSR 20, No. 5 546-51 (1956).
36. Ermolenko, I. and Pavlyuchenko, M. C.A. 15051d (1958) Zhur. Obshchei Khim., 28, 722-8 (1958).
37. Pavlyuchenko, M. et al C.A. 20187g (1960) Zhur, Priklad. Khim. 33, 1385-91 (1960).
38. Kuznetsova, Z.I. et al, Izvest. Akad. Nauk., SSSR, No. 3, 557-59 (1965).
39. Pavlyuchenko, M. et al, C.A. 83, 166024z (1975), Zhur., Priklad. Khim. 48, 1822-5 (1975).
40. Hiatt, G.D. and Crane, C.L. U.S. 2,473,473 (1949).
41. Chú, N.J., Pulp and Paper Inst. of Canada, Report #42, 1970.
42. Williams, H.D., U.S. 3,236,669 (1966).
43. Hergert, H.L. and Zopolis, P., Fr. Pat. 1,469,890, C.A. 68 41234b (1968).
44. Nakao, O., et al, Canadian Patent 876,148 (1971).
45. Nakao, O., et al, U.S. 3,669,916.
46. Mahomed, R.S., B.P. 1,309,234 (1973).
47. Clermont, L.P., Canadian Patent 899,559 (1969).
48. Clermont, L.P. and Bender, F., J. Poly. Sci., 10 (6), 1665-77 A-1 (1972).
49. Venkateswaran, A. and Clermont, P., J. Appl. Poly. Sci., 18, 133-42 (1974).
50. Bender, F., et al, U.S. 3,715,268 (1973).
51. Schweiger, R.G., Chem. & Ind. 296, (1969).
52. Schweiger, R.G., German Patent 2,120,964 (1971).
53. Schweiger, R.G., U.S. 3,702,843 (1972).
54. Schweiger, R.G., TAPPI, 7th Dissolving Pulp Conf., Atlanta (1973).
55. Schweiger, R.G., TAPPI, 57 #1, 86-90, 1974.
56. Schweiger, R.G., J. Org. Chem., 41, (1) 90-93 (1976).
57. Gray, P., Chemical Reviews, 1069, (1955).
58. White, E.H. and Feldman, W.R., J.A.C.S. 79, 5832-33 (1957).
59. Pasteka, M. and Mislovicova, D., Cellulose Chem. & Tech., 8, 107-114 (1974).

60. Ibid, 481-486 (1974).
61. ibid, 9, 325-330 (1975).
62. Grinshpan., O., et al, Daklod, Akad. Nauk, Belloross, SSSR, 18, (9) 828-31 (1974).
63. Grinshpan, D., Kaputskii, F.N., et al, Belloross, Gos. Univ., Minsk., SSSR.; C.A. 194931 (1975).
64. Bashmakov, I.A. et al, Vestsi Akad Nauk, Gos Univ. Belloross SSSR, (4) 29-32 (1973) C.A. 28636u (1973).
65. Nicholson, M. and Johnson, D.C., TAPPI, 8th Dissolving Pulp Conf., Syracuse, N.Y. 1975.
66. Seymour, R.B. and Johnson, E.L., Organic Coatings and Plastics Preprints, ACS Mtg. San Francisco 665-73 (1976).
67. Seymour, R.B. and Johnson, E.L., Polymer Preprints, 17, #2, 382-383 (1976) (ACS, San Francisco Mtg.).
68. Lowe, O.G., J. Org. Chem. 41 (11) 2061-64 (1976).

The Spinning of Unconventional Cellulose Solutions

D. M. MacDONALD

International Paper Co., Tuxedo Park, N.Y. 10987

About ten years ago it became apparent that the viscose process was encountering problems due to high capital costs and environmental effects. Use of an alternative route to regenerated cellulose was indicated and a project to screen the various cellulose solvents was started. In this paper, some of the results will be described.

In the first experiments, Table I, the solutions were prepared using kraft hardwood dissolving pulps. Cellulose, dissolved in 72% sulfuric acid, hydrolyzed to a DP below 30 in less than 15 minutes at $0°C$. Phosphoric acid (85%) gave a very viscous 2% cellulose solution of practical DP, but the presence of gels and fibers made filtration very difficult and H_3PO_4 could not be washed from films cast on glass plates. Neutralization with 14% ammonium hydroxide, followed by washing, gave complete removal of phosphate, but higher concentrations of ammonium hydroxide gave crystalline ammonium phosphate in the film.

Table I

Critique of Cellulose Solvents First Examined

Solvent	Result
72% H_2SO_4	Excessive DP loss
85% H_3PO_4	High gel and fiber, H_3PO_4 hard to remove
SO_2 - NH_3	Good solution, impractical to regenerate
SO_2 - $(CH_3)_2$ NH	Good solution, impractical to regenerate
64% $ZnCl_2$	Good solution, very slow coagulation

The extreme difficulty of filtration along with the poor economics anticipated in using a 2% solution prompted an examination of the newer SO_2-amine solvents reported by Hata and Yokota in Japan (1-3). These solvents gave good solutions but,

when regenerated in water at 0°C or higher, SO_2 evolution produced a foam. As reported by these Japanese workers (3), regeneration at -10°C in methanol gave a clear film, but the
economics of working under pressure at such low temperature
seemed doubtful.

The final solvent in Table I, aqueous zinc chloride, gave an
excellent solution which, unfortunately, could not be coagulated
in a form with strength would permit spinning. This was deduced
from experiments utilizing hand-cast films which showed little or
no cohesion.

Spinning Experiments Using Cellulose Dispersed in Calcium Thiocyanate

While examining the above solvents, flocked pulp dispersed
in aqueous calcium thiocyanate was pressed in a laboratory press
and a clear film was produced. Since fiber extruders are used in
preparing synthetic fibers, it seemed possible that this process,
illustrated in Figure 1, could give a practical route to regenerated cellulose; therefore, a closer investigation of this
unusual solvent system, first described by Bechtold and Weratz
of DuPont (4-6), was commenced.

In the anticipated industrial process, pulp would be
flocked, mixed with calcium thiocyanate solution, drained, and
then pressed. These steps are very similar to the caustic
steeping, pressing, and shredding stages of the viscose process,
so viscose pilot plant equipment was used in our work. The
shredded cellulose calcium thiocyanate fibers would be fed into
an extruder and extruded through a filament die into an aqueous
bath containing calcium thiocyanate.

A Brabender Model 252 Extruder with four heating zones, a
Hastalloy barrel, and a chrome-plated, uniform taper screw of 25
flites with a 3:1 compression ratio was used in the lab studies.
Two eight-hole circular dies (3/4 x 1/8 in.) were used; hole
diameters were .013 in. and .006 in.; capillary length was .010
in.; a 200 x 1400 Dutch Weave stainless steel filter was posi-
tioned behind the die. The micron rating of the filter was 12.

The initial experiments showed that fibers could be extruded, but that these fibers could not be stretched in air even
when the air temperature was raised as high as 180°C, at which
point decomposition of thiocyanate was occurring. Hole blockage
was a recurring problem even though the filter-hole size was far
smaller than the die-hole diameter.

Difficulty was also experienced with feeding the flocked
dispersion to the extruder screw because of blockage at the
hopper neck. This problem was solved by cutting the pulp into
1/4-inch squares after sheet steeping. A problem with variable
extrusion behavior was finally traced to variations in calcium
thiocyanate composition. Consultations with the manufacturer,
Halby Chemical Co., revealed that ammonium thiocyanate was the

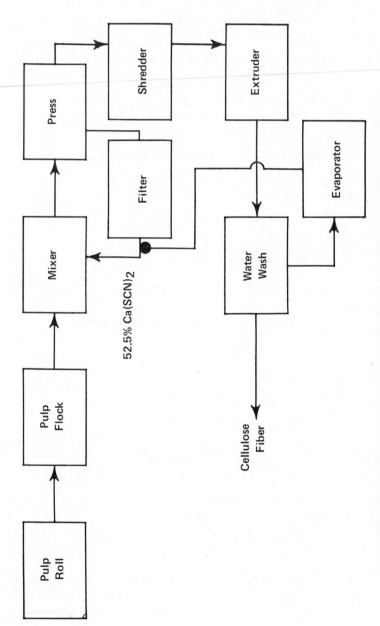

Figure 1:   Flow diagram for rayon production utilizing a cellulose–calcium thiocyanate dispersion

most likely impurity, the addition of ammonium thiocyanate to the calcium thiocyanate solution was examined. About 1% ammonium thiocyanate with 52.5% calcium thiocyanate is optimum; less than 0.2% ammonium thiocyanate gives an unextrudable dispersion.

Table II illustrates what seems to be the optimized process. Calcium thiocyanate concentrations below 45 or 50% do not give a product which can be extruded, and concentrations of ammonium thiocyanate above 1% cause filaments to stick to godets and guides. An alkyl aryl, non-ionic surfactant was used to aid penetration during steeping. The long steeping time seems unnecessary if flock is used rather than sheets; however, the flock did not feed properly as was mentioned previously. Our impression is that flock would very probably feed into a large extruder. Normally, the sheets press easily, but extrusion temperatures above about 120°C cause discoloration of the filaments, apparently due to some thiocyanate decomposition.

Table II

Optimum Conditions for Extrusion of the Cellulose
$Ca(SCN)_2$ Dispersions

Pulp: low viscosity, low resin, dissolving grade

Steeping solution: 52.5% $Ca(SCN)_2$, 0.3-1.0% NH_4SCN, 0.1%
non-ionic surfactant, room temperature,
2 days

Pulp: Solution (after pressing) - 1: 2.2-2.8

Dispersion form: 1/4-inch squares

Extrusion temperature: 110°C (all 4 zones).

Bath: Water, 25% $Ca(SCN)_2$, 25% NaCl.

Using material prepared under these optimum specifications with the .013-in. hole diameter die, filaments could be necked down to .002 in. dry diameter by speeding up the first godet. Stretch between the first and second godets was poor (normally only about 10%) even with 100°C aqueous calcium thiocyanate in the stretch bath. With the .006-in. hole diameter die, filament diameter did not decrease proportionally as swelling at the die face became much more pronounced. This was interpreted as showing the need for lower viscosity and, consequently, a higher pulp: thiocyanate ratio.

Dilution of the dispersion with additional thiocyanate gave a homogeneous mixture if done in a Banbury-type mixer. However, the extruded fibers refused to coagulate properly unless the extruder temperature was reduced to as low as 60°C. Fiber strength was extremely poor.

Finally, addition of DMSO, methanol, ethylene glycol, boric acid, EDTA, hydrazine, and hydroxylamine to the calcium thiocyanate solution was evaluated, but stretchability was not improved. Syntheti̧c polymers were also were added without benefit.

Discussion this far has concerned extrudability of the various dispersions. Deliveries were low (only 30 g/min at top extruder speed), and spinning speeds above 30 m/min were reached only rarely. Most noticeable was an extreme tendency to shrink upon drying. Slack-dried yarn shrinks 30% or more; restrained drying on cones gave very fuzzy yarns, due to filament breakage.

Typical yarn properties illustrated in Table III were very poor, with most runs giving about 0.3 g/den conditioned tenacity (although some yarns have given 0.8 g/den). Elongations were less than 10%, with values up to 15% being observed. Very noticeable was the tendency for all yarns to break at about the same stress, therefore the lowest deniers always gave the strongest yarn.   Yarn of still lower denier, however, could not be spun with available extrusion dies. Because of swelling at the die face, it is unlikely that smaller diameter holes, which are very difficult to drill, would help. The die must also be quite thick as pressures of 500–4000 lb/in.^{2} have been observed.

Table III

Typical Rayon Properties from the Thiocyanate Cellulose Dispersions

Cond. Ten., g/den	–	0.1–0.8
Cond. Elong., %	–	4–10%
Wet Ten., g/den	–	0.1–0.3
Wet Elong, %	–	10–20%

Spinning Experiments Utilizing Cellulose Dissolved in NO_2-DMF

Turning now to the nitrogen dioxide–DMF solvent, results indicate that in so far as solution properties and ease of solution preparation are concerned, this is an excellent combination that produces yarn properties better than those of the thiocyanate cellulose–spun yarn.

Our research was prompted by a paper presented by Schweiger (5) at the TAPPI Dissolving Pulp Conference in Atlanta. In this paper, Schweiger described both the facile reaction of cellulose with nitrogen dioxide in dimethylformamide solution (NO_2-DMF) to form a cellulose trinitrite solution, and the almost instantaneous regeneration of cellulose when this solution is added to water.

$$CellOH + NO_2 \xrightarrow[DMF]{} CellONO + HNO_3$$

$$\downarrow H_2O$$

$$CellOH + HNO_2$$

Work was initiated to investigate the belief that production of rayon from cellulose spun in NO_2-DMF might be feasible if the plant was adjacent to an ammonium nitrate plant supplying NO_2 and ammonia, and which could process the ammonium nitrate by-product. The tonnage of ammonium nitrate would be high – about 3 times more ammonium nitrate would be produced than cellulose. A flow diagram for a hypothetical production unit is given in Figure 2.

With this process, pulp flock (7-8% moisture) would be treated with about 2/3 of the required quantity of DMF and, after about two hours, this would be transferred to a reactor to be combined with six moles of NO_2 (corrected for pulp moisture) and the remaining DMF. The solution forms at room temperature after 10-30 minutes and would then be filtered, with simultaneous deaeration if a vacuum filter is used. Spinning is into a bath containing water, DMF, and ammonium nitrate at pH 7 to minimize DMF hydrolysis. DMF is recovered from excess spin bath, either by fractional distillation or methylene dichloride extraction, and is returned to the DMF reservoir. The residual ammonium nitrate solution would be sent for recovery of solid ammonium nitrate.

Conventional viscose processing equipment was used in the experiments. The pulp, usually a low viscosity, hardwood kraft of dissolving grade, was first flocked and placed in a viscose mixing can with 2/3 of the desired DMF. After two hours, the required amount of NO_2 was added in the remaining DMF (2.5 parts NO_2 per part of cellulose was usually used) and mixing, with the can partially submerged in a $2°C$ bath, was started. After 30 minutes (10 minutes seemed sufficient), the can was removed from the bath, a standpipe was inserted, and the solution was forced by application of 50 psi air pressure to pass through a 6-micron polypropylene filter into an adjacent, evacuated can. Usually the vacuum was applied for about 2 hours after the end of filtration to complete deaeration; whether this is necessary is doubtful.

Every time a can was opened and exposed to the air, a few drops of liquid NO_2 were added. This prevents the skin formation which occurs rapidly in air, nitrogen, or even vacuum. Apparently, NO_2 gas effectively suppresses reversal of the nitrite formation reaction.

Cellulose DP slowly drifts downward on storage of the solution, the best data on this effect are included in a paper by Pasteka (8). However, many times we have been able to spin successfully after storing the solution overnight. It is im-

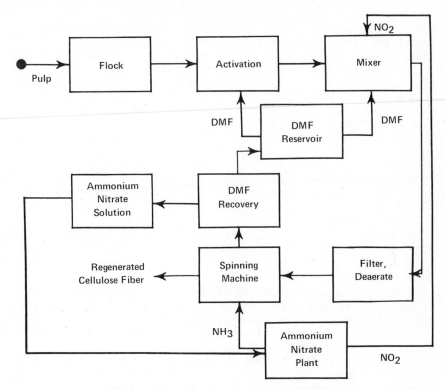

Figure 2:   Flow diagram for NO₂–DMF spinning

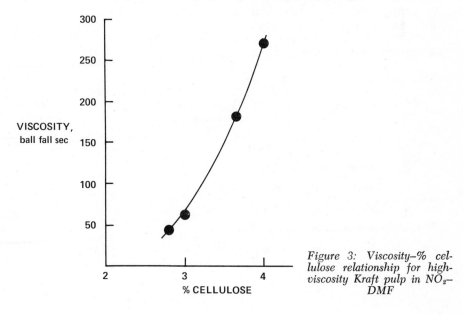

Figure 3: Viscosity–% cellulose relationship for high-viscosity Kraft pulp in NO₂–DMF

portant that the solutions be stored at room temperature, or
lower, and that temperatures above 30°C be avoided during dissolution. Frictional heat occurs, and efficient cooling during
stirring is necessary. As Schweiger first demonstrated, nitration occurs due to the transesterification with by-product nitric
acid at temperatures above room temperature. The data in Table
IV show that less nitration occurs when a pulp of normal moisture
is used in place of a bone dry pulp.

Table IV

% Cellulose-Bound Nitrogen after Regeneration Compared
to Solution Age and Pulp Moisture

Pulp Moisture %	% N 1 hr	% N 24 hr
0	0.09	0.50
7	0.03	0.18

Solution viscosity was rather high, as is illustrated in
Figure 3, where a high viscosity pulp was used. Low viscosity
pulp will cause the curve to shift about 0.5% to the left.

Spinning was through a 720-hole, .0025-in. hole diameter
platinium spinneret. Hole blockage was never observed.

In the next tables, some strength results are compared with
spin bath composition. In Table V, a few baths which did not
give successful spinning are described. Included here is methylene dichloride, which is a low boiling solvent capable of
extracting DMF from water. Methylene dichloride boiled at the
spinneret face thereby indicating a high heat evolution. Also
weak or strong acids, and strong bases seem harmful, while DMF-
NH_4NO_3 mixtures do not give sufficiently fast coagulation.

Table V

Spinbaths found not Suitable for Spinning
the Cellulose - NO_2 - DMF Solutions

1.   Methylene dichloride

2.   Methylene dichloride - DMF - water

3.   Acetic acid - DMF - water

4.   50 and 75% methanol in DMF

5.   Rayon-type spin baths

6.   Systems cont. NaOH in spin, or stretch, baths

7.   DMF - NH_4NO_3

The addition of water into DMF/ammonium nitrate mixtures allows spinning to proceed normally and yarn properties were unchanged for H_2O, NH_4NO_3-H_2O and DMF-NH_4NO_3-H_2O spinbaths. The results in Table VI are typical: strength values were a bit below ordinary rayon and elongation was low when 3% solutions were spun at 10 m/min. Low elongation indicates a yarn which is too brittle for easy textile processing.

Table VI

Effect of DMF and Ammonium Nitrate in Aqueous
Spin Baths at 24°C (3% cellulose, 10m/min)

Bath Comp. (%)*	Cond. Ten. (g/den.)	Cond. Elong. (%)	Wet Ten. (g/den.)	Wet Elong. (%)
H_2O	2.0	4.9	0.8	12
33N	2.1	3.4	0.8	6.3
44D,37N	1.8	2.1	0.6	3.8

*D = DMF, N = NH_4NO_3

Table VII shows a few experiments where 5% solutions were spun at 30 m/min into spin baths containing varying quantities of DMF-NH_4NO_3 and water. Again, the tensile properties were not good.

Table VII

Effect of DMF and Ammonium Nitrate in Aqueous
Spin Baths at 24°C (5% cellulose, 30 m/min)

Bath Comp. (%)*	Cond. Ten. (g/den.)	Cond. Elong. (%)	Wet Ten. (g/den.)	Wet Elong. (%)
41D, 29N	1.7	2.9	0.7	7.5
30D, 26N	1.7	2.9	0.6	9.2
28D, 19N	1.7	2.7	0.7	7.8
16D, 11N	1.7	3.0	0.7	9.5

*D = DMF, N = NH_4NO_3

Use of divalent calcium and trivalent aluminum was evaluated as an aid to coagulation; yarn properties were unchanged. Table VIII gives results for $24°$ and $50°C$ spin bath temperature. The $50°C$ yarn strength results were very slightly better than the $24°C$ results.

Table VIII

Effect of Aluminum and Calcium Nitrates on Spinning
6% Solutions at 40 m/min

Bath Comp. (%)*	Cond. Ten. (g/den.)	Cond. Elong. (%)	Wet. Ten. (g/den.)	Wet. Elong. (%)
7A, 23N (24°C)	1.6	4.0	0.6	8.6
7C, 23N (24°C)	1.6	4.0	0.5	8.5
14A, 17N (24°C)	1.7	4.7	0.6	10.5
14C, 17N (24°C)	1.6	4.9	0.5	11.0
7A, 23N (50°C)	2.0	5.0	0.7	9.8
7C, 23N (50°C)	1.8	4.2	0.6	9.8
14A, 17N (50°C)	1.7	4.0	0.6	8.9
14C, 17N (50°C)	2.1	5.3	0.8	10.3

*A = $Al(NO_3)_3 \cdot 9H_2O$; C = $Ca(NO_3)_2 \cdot 4H_2O$; N = NH_4NO_3

Spinning speed, % cellulose in the NO_2/DMF solution, and the presence of glycerol in the stretch bath gave the yarn properties illustrated in Table IX. Water at $95°C$ seemed best for the stretch bath in previous experiments. Spinning speeds were 10-60 meters/ min and the range of 3% to 7% cellulose was examined.

Table IX

Effect of Spinning Speed, Glycerol in Stretch Bath
and % Cellulose on Spinning into water at $25°C$

			Yarn Properties			
% Cell	Stretch Bath Comp.	Spinning Speed m/min	Cond. Ten. g/den	Cond. Elong. (%)	Wet Ten. (g/den)	Wet Elong. (%)
3	H_2O, 95°C	10	1.8	3.1	0.78	7.1
3	50% glycerol 92°C	10	1.8	2.3	0.81	6.3
5	H_2O, 95°C	30	1.8	3.7	0.76	7.1
5	H_2O, 95°C	40	1.7	4.5	0.72	8.0
5	H_2O, 95°C	60	1.6	6.0	0.60	9.5
7	H_2O, 95°C	30	1.8	3.3	0.78	6.0

A sodium alkyl sulfonate wetting agent in the spin bath did not improve the results, shown in Table X; however, higher spin bath temperatures (to 50°C) again gave a very slight strength improvement without improved elongation.

Table X

Effect of Spin Bath Wetting Agent and Aqueous Spin Bath
Temperature on Acetakraft Spinning at 30 m/min

% Cell.	Spin Bath Comp. and Temp.	Cond. Ten. g/den.	Yarn Properties Cond. Elong. (%)	Wet Ten. g/den.	Wet Elong. (%)
7	H_2O, 23°C	1.8	3.3	0.78	6.0
7	H_2O + 1% wetting agent, 23°C	2.1	4.2	0.80	6.3
7	H_2O + 1% wetting agent, 30°C	2.2	4.0	0.88	7.9
7	H_2O + 1% wetting agent, 45°C	2.2	4.1	0.94	15.4
3	H_2O, 50°C	2.2	4.5	0.73	7.0

Perhaps the most striking thing about these results is the lack of response to widely varying spinbath compositions; in fact, the same yarn properties were found for spinning into methanol. This prompted a brief look at fiber electron micrographs. Gas bubbles, probably from decomposition of nitrous acid to nitric oxide, NO, have caused fiber damage as shown by the presence of a hollow core in Figure 4. Also, the fibers have stuck together indicating that the coagulation rate is too slow.

Figure 5 gives a side view illustrating a bundle of fibers where one fiber wall has been ruptured by a bubble. As non-aqueous spin baths were unsuccessful and coagulation in presence of water was insufficiently fast to prevent fiber sticking and nitric oxide damage, it was concluded that the NO_2-DMF solvent had little chance of industrial success in the Figure 2 process.

Spinning of Cellulose in DMSO – Paraformaldehyde Solutions

Fibers have also been successfully spun from solutions of cellulose in DMSO-PF. This interesting solvent system was first described by D.C. Johnson and coworkers (9, 10). Cellulose is suspended in DMSO and the mixture heated to about 120°C; solid paraformaldehyde (PF) is then added. Monomeric formaldehyde smoothly forms and reacts with cellulose to form a methylol derivative. This derivative is stable and soluble in DMSO.

Figure 4: Electron micrograph of NO$_2$–DMF fibers showing hollow cores and adhering fibers

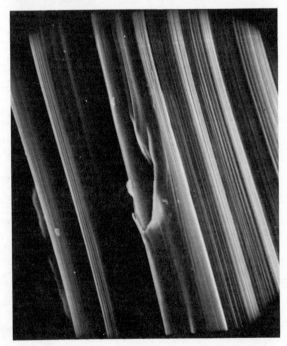

Figure 5: Electron micrograph of NO$_2$–DMF fibers showing wall rupture by a gas bubble

$$(CH_2O)_x \longrightarrow XCH_2O$$

$$CH_2O + CellOH \xrightarrow[DMSO]{} CellOCH_2OH$$
$$\downarrow H^+ \text{ or }$$
$$\downarrow OH^-$$
$$CellOH$$

If the solution is poured into water, coagulation occurs. Speed of coagulation is too slow for successful spinning into water. The same is true for acetone, benzene, acetic acid, acetone-water and acetic acid-water. As methylol decomposition is catalysed by either strong acid or strong base, acids and bases were evaluated next: fibers have been successfully spun from both acidic and basic baths.

MC&B paraformaldehyde and reagent grade DMSO were used. The 6% solutions were prepared in a 4-1 beaker using the conditions described by Johnson, and were filtered in the equipment previously described for NO_2-DMF solutions. While filtration and deaeration did not cause problems, the PF must be added very slowly during solution preparation or the mixture will cool due to the endothermic nature of the process and the solution will not form. Fiber spinning did not take place in water, in 0.5% and 1% H_2SO_4, or in 8% and 20% Na_2SO_4, as shown in Table XI.

Table XI

Unsuccessful Spin Baths for Cellulose
Regeneration from DMSO - PF Solutions

H_2O

0.5% H_2SO_4

1% H_2SO_4

8% H_2SO_4, 20% Na_2SO_4

Fibers were spun with difficulty in 1% NaOH (Table XII). The yarn was very weak between the spinneret and the first godet, and frequent breaks occurred. Yarn test properties were very similar to NO_2-DMF yarns and brittleness is illustrated by the low percent elongations. Spinning was greatly improved by use of high NaCl in presence of NaOH. The yarn was still brittle.

Table XII

Successful Spin baths and Yarn Properties for Cellulose
Regenerated for DMSO-PF Solutions

Bath Composition and temperature	Yarn Properties			
	Cond. Ten,g/den	Cond. Elong, %	Wet Ten,g/den	Wet Elong,%
1%, NaOH, 55°	2.2	4.8	0.8	7.5
1%, NaOH, 24% NaCl, 55°	1.9	5.4	1.0	8.1
10% H, 4.2% Z, 15.2% N, 25°C	2.6	4.4	1.2	5.1
10% H, 4.2% Z, 15.2% N, 50°C	1.9	9.8	0.7	18.9
10% H, 4.2% Z, 15.2% N, 70°C	1.3	9.3	0.5	21.6

$H = H_2SO_4$, $Z = ZnSO_4$, $N = Na_2SO_4$

A high acid, rayon-type spin bath containing zinc sulfate gave good spinning with 30-40% stretch in a 95°C aqueous stretch bath. Spinning speed was only 6 meters/min; this was dictated by the volume of solution available and there does not seem to be any reason why spinning would not succeed at higher speeds.

Yarn properties were still below commercial acceptability; however, yarn properties did vary with spin bath temperature and this was gratifying in view of the independence of NO_2-DMF yarn properties with regard to spinning conditions. Yarn tenacity falls as temperature is increased, while elongation increases with temperature. At 50°C, results were quite close to commercial acceptability.

Microscopic examination of the yarns showed clumps of mutually adherent fibers in all yarns, except for the sample spun in 70°C acid. Thus, stronger coagulation conditions are strongly indicated. Higher acid and zinc sulfate, or replacement of zinc sulfate with the environmentally more desirable aluminum sulfate, should be tried at 40 or 50°C. Unfortunately, this experiment has not yet been carried out.

CONCLUSIONS

(a) Industrial utilization for rayon production is doubtful for cellulose dispersed in strong aqueous calcium thiocyanate.

(b) Cellulose-NO_2-DMF solutions are more promising although use is doubtful in a combined ammonium nitrate-rayon process.

(c) Cellulose-DMSO-PF solutions are also promising; real difficulty can be foreseen in devising an efficient recovery system for DMSO, formaldehyde, and the spin bath components.

LITERATURE CITED

1. Hata, K. and Yokota, K., Sen-i Gakkaiski, (1966) $\underline{22}$ (2), 96-102.
2. Hata, K. and Yokota, K., Sen-i Gakkaiski, (1968) $\underline{24}$ (9), 415-419.
3. Hata, K. and Yokota, K., Sen-i Gakkaiski, (1968) $\underline{24}$ (9), 420-424.
4. Bechtold, M.F. and Werntz, J.H., U.S. 2,737,437 dated March 6, 1956.
5. Bechtold, M.F. and Werntz, J.H., U.S. 2,737,459 dated March 6, 1956.
6. Bechtold, M.F. and Werntz, J.H., U.S. 2,810,162 dated October 22, 1957.
7. Schweiger, R.G., TAPPI, (1974) $\underline{57}$, 86-90.
8. Pasteka, M. and Mislovicova D., Cellulose Chem. Tech., (1974) $\underline{8}$, 107-114.
9. Johnson, D.C., Nicholson, M.D. and Haigh, F.C., J. Appl. Polymer Sci., Appl. Polymer Symposia, (1976) $\underline{28}$, 931-943.
10. Swenson, H.A., J. Appl. Polymer Sci., Appl. Polymer Symposia, (1976) $\underline{28}$, 945-952.

4

Production of Rayon from Solutions of Cellulose in N₂O₄–DMF

R. B. HAMMER and A. F. TURBAK

ITT Rayonier Inc., Eastern Research Div., Whippany, N.J. 07981

A wide range of wood pulps in various physical forms were
found to dissolve readily in a combination of DMF/N_2O_4. The concentration of pulp dissolved was a direct function of the degree
of polymerization. Dimethylformamide was compared to dimethyl
sulfoxide and acetonitrile and was found to be preferable overall
with respect to solvent power, viscosity of solutions, stability
and recovery. The temperature of N_2O_4 addition and the resultant
time for dissolution were found to be critically related to ultimate fiber physical and analytical properties. Fibers with a
high wet modulus and intermediate tenacity were readily produced
from proton donor systems involving hydroxylic coagulation baths
such as water, alcohols and glycols. A wide variety of fiber
cross sections could be produced and proved to be related to the
nature of the regenerant employed during spinning.

Introduction

The most widely used method for converting wood pulp into
regenerated cellulose for films and fiber production is the viscose process. However, there are several problems associated
with the viscose process which do not appear to be diminishing
even in view of current developments. There are other processes
for producing regenerated cellulosic products including regeneration from cellulose nitrate which is very hazardous, and cuprammonium hydroxide which forms a soluble cellulose complex.
However, the production of cellulosic articles from these processes is minuscule compared to the viscose method.

There are entirely different classes of chemical systems
which are non-aqueous which dissolve cellulose. Dinitrogen tetroxide, N_2O_4, has been used as an agent to make 6-carboxy cellulose in non-polar solvents such as chloroform or carbon tetra-
chloride. In 1947-48 W.F. Fowler et al[1] evaluated the relative
solvent power of forty-five organic solvents with N_2O_4 for

dissolving cellulose. In separate patents, H.D. Williams (2) and
H.L. Hergert (3) et al reported dimethyl sulfoxide as a specific
polar solvent for use with N_2O_4 to dissolve cellulose.

In later work, the solution properties of cellulose/N_2O_4/DMF
solutions or cellulose plus other   polymer/N_2O_4/DMF solutions
were examined. For example, O. Nakao reported graft copolymers
prepared from cellulose/N_2O_4 solutions.(4) L.P. Clermont(5) and
R.G. Schweiger(6) reported that certain cellulose derivatives
could be prepared through the use of cellulose/N_2O_4/DMF solutions.
Basic chemical and physical studies on this system have been reported by N.J. Chu(7) and M. Pasteka and D. Mislovicova.(8)

The experiments to be described were intended to explore the
use of cellulose solutions in N_2O_4/DMF or DMSO for the production
of regenerated fibers as a potential replacement for the viscose
process.

Experimental

Experiments were designed to determine the influence of the
form of the pulp and the degree of polymerization prior to dissolution. Although the majority of the solutions were prepared
using Abbe'cut material, i.e. a highly powdered, defibered pulp,
the dissolution process is not limited with respect to the degree
of polymerization or the pulp form.

All solution compositions are given as weight percents in
the order pulp/N_2O_4/solvent, eg. 8/15/77 represents 8% cellulose,
15% N_2O_4 and 77% solvent.

A wide variety of cellulosic pulps were found to dissolve in
the N_2O_4 solvent system. Both sulfate and sulfite pulps can
readily be employed and pulps currently available for the viscose
process are among those that may be used. In addition, bleached
or non-bleached, barked and non-debarked pulp samples also
readily dissolve in this system. These pulps may consist of
hardwoods, softwoods or mixtures of the two species. A typical
example of the solution preparation procedure is described below.

Silvanier-J, a prehydrolyzed kraft pulp of 1050 D.P., after
converting to alkali cellulose by methods well known in the rayon
industry, was alkaline aged to a D.P. level of 450, neutralized,
washed, dried, then either fluffed, diced or defibered.

An 8/15/77 cellulose/N_2O_4/DMF solution was prepared by
charging 160 parts of this alkali aged prehydrolyzed, kraft pulp
(D.P. 450) and 1540 parts of dimethylformamide (DMF) into a two-
liter four neck resin reaction flask equipped with a stainless-
steel mechanical stirrer, thermometer, and a 250 ml. equalizing
pressure addition funnel. The resulting slurry was stirred and
cooled to below +20°C, preferably between -5°C and +10°C, while
300 parts of liquid nitrogen tetroxide (N_2O_4) was added dropwise
over ca. 60 minute time period. The temperature of the resulting
exothermic reaction was maintained below 20°C, preferably in the
range previously specified during N_2O_4 addition and for the

duration of the remaining dissolution process.

Cellulose/N_2O_4/CH_3CN solutions of 8/20/72 composition were prepared in a similar manner.

Cellulose/N_2O_4/DMSO solutions of 8/15/77 composition were prepared under similar conditions except that the liquid N_2O_4 was first added to the DMSO containing 1.5 parts of water. The cellulose was then added to the cooled (20°C) N_2O_4/DMSO mixture.

All solutions were observed microscopically to be free of gels and unreacted fibers. The solutions were filtered through a 90 mm. diameter nylon, in-line filter during spinning. The solutions were deaerated prior to spinning and viscosities measured by a Brookfield Viscometer and found to be in the range of 8,000-16,000 cps. at 22°C.

The majority of the spinning trials were performed on a bench-scale vertical spinning unit. The solutions were spun into the appropriate primary regeneration bath and the resulting fibers passed vertically to a primary godet, then through a secondary bath to a secondary godet, whose speed could be altered to produce desired stretch conditions.

Several types of spinnerettes have been used successfully to spin fibers from these solvent systems. For example, gold-platinum, typical for the viscose process, stainless-steel, typical for cellulose acetate spinning, and glass. The glass spinnerettes have afforded the best results and are thus preferred since they are both inexpensive and offer the advantage of larger hole length/diameter ratios.

Results and Discussion

The dissolution of cellulose in N_2O_4/solvent systems is believed to result from the formation of a cellulose nitrite ester and nitric acid. (9) Consequently, the regeneration of cellulose during spinning from a true cellulose nitrite ester would require a transnitrosation reaction by some agent to remove the nitrite and provide a hydrogen ion to cellulose. These requirements are met by protonic nucleophilic species such as water, alcohol, and others. If the regenerant-coagulant were water or alcohol then the spent regeneration bath would contain nitrous acid, HNO_2, and/or the alkyl nitrite, RONO, in addition to the DMF and HNO_3 introduced by the cellulose solution. Any unreacted N_2O_4 in the cellulose solution would result in the formation of additional HNO_2 (or RONO) and HNO_3.

Because of the rapidity of coagulation and regeneration of the cellulose from the N_2O_4 solutions, basic fiber properties are determined to a large extent by the composition of the regeneration bath and the spinning conditions employed immediately during and after extrusion. Therefore, initial jet stretch is important in determining the overall fiber physical properties. As regeneration is retarded, godet to godet stretch becomes more important and more effective but is still limited by the initial jet stretch. For example, regeneration can be retarded by the addition of bases such as pyridine to the cellulose solution to neutralize the nitric acid present. In this manner, a cellulose nitrite fiber can be obtained which initially was observed to redissolve in dimethylformamide.

A wide range of solvents may be used to regenerate the solution emerging from the spinnerette. Virtually any liquid which is compatible with and causes extraction of the organic solvent and further de-esterifies the cellulose may be employed. A few of the solvents which are applicable to coagulation include water, a wide range of alcohols, ethylene glycol and mixtures of these or other protic solvents with dimethylformamide. In some cases, spinning and the resulting fiber physical properties may be aided or improved by the presence of soluble salts or bases in the coagulation bath. However, for such a process to be commercially feasible, such compositions would have to be balanced between fiber properties and the economy of the recovery and recycle process for the chemicals involved.

By varying the composition of the regeneration or spin bath, it is possible to obtain a wide variety of fiber properties, from those of regular rayon staple to medium-high performance rayon. For example, use of lower molecular weight alcohols such as methanol or ethanol affords fibers with good tensile properties including wet modulus. The addition of soluble bases to such regenerants appears to improve wet strength while lowering $S_{6.5}$ and water retention values. The $S_{6.5}$ values are an indication of a regenerated cellulosic fiber's solubility in 6.5% sodium hydroxide at 20°C. This is a useful test for determining the potential resistance of such fibers or resultant fabrics to alkaline treatment such as alkaline laundering or mercerization. Accordingly, regular viscose rayon which cannot be mercerized and is not resistant to alkaline washing, unless crosslinked, has a relatively high $S_{6.5}$ of from 25-35%. On the other hand, the high performance and polynosic rayons have superior resistance to caustic soda as evidenced by $S_{6.5}$ values of from 5-15%.

We have recently found that in addition to using soluble bases in the primary regeneration bath, that this important fiber property can be obtained by careful control of the conditions employed during solution preparation ie, the temperature during N_2O_4 addition and the resultant time and temperature for total dissolution. Our studies have definitely established that NO_2 oxidation should be minimized so that low temperatures of N_2O_4 addition (below 10°C for example with DMF) and short dissolution times (4 hours at less than 20°C) are mandatory if $S_{6.5}$ levels are to be maintained in a reasonable range (10-30%) for commercial usefulness of the resulting rayon fibers.

It should be noted further that in this spinning system it is readily possible to obtain fibers with high wet modulus without the use of zinc or other additives which are required in a viscose spinning operation. In addition, fibers with very fine deniers can be readily produced, eg. 0.5 denier, while maintain-

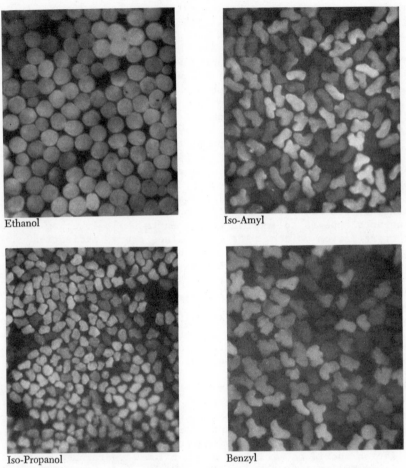

Ethanol

Iso-Amyl

Iso-Propanol

Benzyl

Figure 1. Fiber cross sections resulting from various coagulants (430X)

ing reasonable tensile strength. Although optimum conditions were not established, successful spinning trials were easily performed at 115 meters/minute; faster speeds apparently being limited only by equipment and possibly by liquid flow characteristics, but not by kinetics of regeneration.

Variations in the type of regenerants results in fibers with different cross-sections. Water for example gives a serrated, gear-shaped cross-section; methanol yields a circular cross-section; isopropanol affords irregular shapes; isoamyl alcohol gives a trilobal cross-section; benzyl alcohol gives x-shaped structures and octanol affords hollow fibers. Several examples of these typical cross-sections obtained from cellulose/N_2O_4/DMF solutions are illustrated in Figures 1 and 2 along with those of Fiber 40 and regular rayon. Similar fiber cross-sections result from cellulose/(N_2O_4)/DMSO solutions.

Even the type of hollow filaments obtained from an octanol regeneration bath appears to be controllable. Filaments with segments resembling bamboo are obtainable as well as those with various lengths of hollow lumens in the center of the fiber. The frequency of the segmentation appears controllable, with segments occurring 5 per inch or 25 per inch depending upon the combination of conditions employed.

In Figure 4 stress-strain curves [conditioned (C) and wet (W)] for an alkaline methanol-spun fiber are shown for comparison with regular and high-wet modulus rayon. The shape of the wet curve is of particular importance since the yielding portions is different from that of other rayons. Low elongation in the fiber results in the steep slope of the conditioned stress-strain curve for these fibers and is an indication of stiffness. If the conditioned elongation could be increased to 10-12% or the slope manipulated to fall between that of high-wet modulus rayon and cotton for example, a superior cellulose fiber may be obtainable. In fact, elongations approaching 10% were observed for fibers spun from pyridine-stabilized cellulose/N_2O_4/DMF solution and from runs employing low jet stretch.

Some typical physical property data are shown in Table I employing a variety of primary bath coagulants for the production of rayon fibers from spinning solutions of 8/15/77 cellulose/N_2O_4/DMF composition. The wide range of properties obtained by varying the chemical composition of the regeneration bath affords an appreciation of the versatility of the system.

Some physical property data are shown in Table II comparing some commercial rayons with those produced from spinning 8/15/77 cellulose/N_2O_4/DMF solutions into an isopropanol primary regeneration bath.

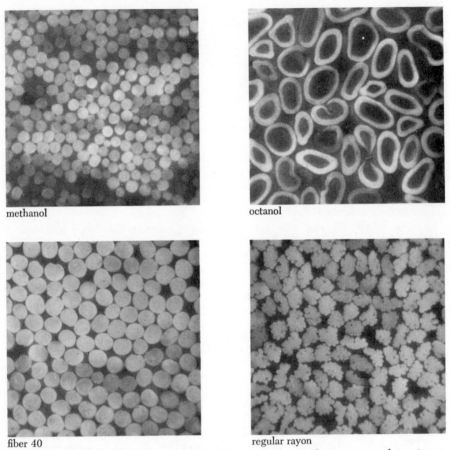

methanol

octanol

fiber 40

regular rayon

Figure 2. Fiber cross sections resulting from various coagulants compared to viscose rayon (430×)

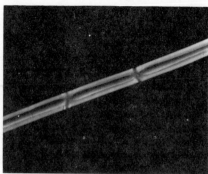

Figure 3. Segmented hollow fiber

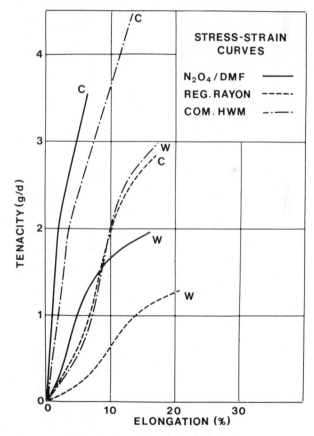

Figure 4. Stress-strain curves of regular rayon, commercial high wet modulus rayon and rayon produced by regenerating cellulose–N_2O_4–DMF solutions from alkaline methanol

TABLE I

PROPERTIES OF FIBERS PREPARED FROM AN 8/15/77 CELLULOSE/N$_2$O$_4$/DMF SOLUTION

Coagulant	Denier	Tenacity, g/d		Elongation, %		Wet Modulus, g/d
		Cond.	Wet	Cond.	Wet	
Isopropanol	0.80	3.58	2.26	6.7	12.9	0.84
Methanol	0.81	3.55	1.46	6.7	13.5	0.73
Water	3.5	2.44	1.30	11.7	15.5	0.51
Octanol	1.6	2.29	1.23	13.2	19.9	0.37
Methanol/1.5% Alkoxide	1.25	2.48	1.51	9.4	18.9	0.49
Isopropanol/1.5% Alkoxide	0.87	3.47	2.13	6.3	11.9	0.98
Methanol/10% Alkoxide	1.8	3.29	1.90	8.4	13.0	1.19

TABLE II

FIBER PHYSICAL PROPERTIES

Staple Type	Tenacity, g/d		Elongation, %		Wet Modulus g/d	% $S_{6.5}$
	Cond.	Wet	Cond.	Wet		
Regular Rayon	1.5-2.8	1.0-1.8	14-25	18-35	0.18-0.28	20-35
Intermediate Wet Modulus Rayon	3.5-5.0	2.5-3.5	12-19	18-24	0.45-0.60	15-20
High Wet Modulus Rayon	3.5-8.0	2.5-6.0	6-14	9-18	0.7-3.0	5-10
8/15/77 Cellulose/N_2O_4/DMF Fiber	2.0-3.6	1.5-2.4	6-19	10-15	0.7-1.7	20-80

Conclusions

A wide range of pulps have been found to readily dissolve in the N_2O_4/DMF system including non-debarked experimental samples. However, groundwood furnish, such as is employed for news print will not dissolve in this system. The pulps may be used in either a fluffed, Abbe'cut, shredded or diced sheet form without encountering dissolution problems.

The concentration of pulp which can be used depends upon the degree of polymerization. At 1000 D.P., concentrations up to 3% can be spun while at 400-500 D.P. up to 8% solutions were feasible, while at 300 D.P. up to 10% cellulose solutions could be readily processed.

Dimethyl formamide was found to be preferable to dimethyl sulfoxide or acetonitrile as the solvent system based upon over-all solution properties. The DMF/N_2O_4 solutions were clear and essentially free from gels or fibers therefore requiring only a single stage polishing filtration prior to spinning.

The temperature of the N_2O_4 addition and the time of dis-solution were found to be critically related to the final $S_{6.5}$ fiber levels. For example, by maintaining the temperature below 20°C, the $S_{6.5}$ level can be held in the 27-30% range which is similar to that of regular rayon.

Coagulation and regeneration of the cellulose/N_2O_4/DMF solutions is extremely rapid in the presence of proton donor systems. Fibers of excellent high wet modulus eg. up to 1.7 g/d can be spun without the need for spin-bath additives.

Spinning speeds of 115 meters per minute were achieved employing only a 3 inch primary bath coagulation travel length. No attempts were made to optimize fiber physical properties at this higher spinning speed, which was the maximum speed obtainable and was limited by the godet size and motor drive units.

Conditioned fiber stretch generally was between 7% and 10% which is normally too low for processing. However, this level could be improved by proper control of jet stretch to godet stretch ratios. Fiber cross sectional shapes can be controlled by the use of various molecular weight alcohols to give either round, serrated, "X" or "Y" shapes or even segmented hollow fibers. These fibers display excellent cover power as compared to regular HWM rayon in knits.

Acknowledgements

We are grateful to Dr. Arthur C. West* who performed much of the initial basic research and experimentation involved in this investigation.

* Present Address - 3-M Company, St. Paul, Minnesota

Literature Cited

1. Fowler, W. F. and Kenyon, W. O., J. Amer. Chem. Soc., <u>69</u> 1636 (1947).
2. Williams, H. D., U.S. Patent No. 3,236,669 (1966).
3. Hergert, H. L. and Zopolis, P. N., French Patent No. 1,469,890 (1967).
4. a) Nakao, O. et. al Canadian Patent No. 876,148 (1971),
   b) U.S. Patent No. 3,669,916 (1972).
5. Clermont, L. P., (a) Canadian Patent No. 899,559 (1972); (b) Monthly Research Notes, Dept. of Fishery and Forestry, Canada <u>26</u>, No. 6, 58 (1970); (c) J. Poly. Sci., <u>10</u>, 1669 (1972); (d) J. Appl. Poly. Sci., <u>18</u>, 133 (1974).
6. Schweiger, R. G., (a) Chemistry and Industry 296 (1969); (b) U.S. Patent No. 3,702,843 (1972); (c) German Patent No. 2,120,964 (1971).
7. Chu, N. J., Pulp and Paper Research Institute of Canada, Report No. 42 (1970).
8. Pasteka, M. and Mislovicova, D., Cellulose Chemistry and Technology <u>8</u>, 107 (1974).
9. Portnoy, N. A., Unpublished Results.

5

Chemistry of the Cellulose–N₂O₄–DMF Solution: Recovery and Recycle of Raw Materials

NORMAN A. PORTNOY and DAVID P. ANDERSON

ITT Rayonier, Inc., Eastern Research Div., Whippany, N.J. 07981

The purpose of the work presented in this paper is to establish a foundation for a technically viable recovery and recycle system based on spinning rayon fibers from the cellulose/N_2O_4/DMF solution. Other papers in this symposium discussed solution makeup, spinning, and fiber properties. The paper presented here, which is divided into two parts, attempts to explain some of the chemistry of dissolution and regeneration as well as our efforts in developing a technically feasible recovery and recycle system.

A. The Chemistry of The Cellulose/N_2O_4/DMF Solution

Introduction

When the investigation of this system was initiated, the importance of delineating the exact mode of dissolution of the cellulose in DMF/N_2O_4 was recognized. To understand the dissolution step and to properly formulate a recovery and recycle system, a thorough study of the chemistry of cellulose in DMF/N_2O_4 had to be undertaken.

From a theoretical viewpoint, several possibilities as to the mechanism of cellulose dissolution in DMF/N_2O_4 are evident. The simplest would be a reaction between the hydroxyl groups of cellulose and N_2O_4 to form a cellulose nitrite ester and HNO_3 as shown in equation 1 (mechanism 1). This cellulose nitrite ester may then be soluble in DMF.

$$1.) \quad CellOH + N_2O_4 \xrightarrow{DMF} CellONO + HNO_3 \xrightarrow{DMF} solution.$$

The components of such a solution would be cellulose nitrite, nitric acid (HNO_3), unreacted N_2O_4, and DMF. This kind of reaction between alcohols and N_2O_4 is well known and has been extensively documented in the literature for a wide range of alcohols. Since cellulose is an alcohol it should react in the same manner. A second possibility is that a complex, probably of the donor-acceptor type, between cellulose and N_2O_4 forms and that DMF acts as a solvent for this complex as in equation 2 (mechanism 2).

$$2.)\ CellOH + N_2O_4 \xrightarrow{\hspace{1cm}} CellOH : N_2O_4 \xrightarrow{\text{DMF}} \text{solution}$$

In this case the spinning solution would contain the cellulose: N_2O_4 complex, uncomplexed N_2O_4 and DMF. A third possibility is that DMF and N_2O_4 form a donor-acceptor complex in which cellulose is soluble as in equations 3a and 3b (mechanism 3).

$$eq.\ 3a.)\quad DMF + N_2O_4 \xrightarrow{\hspace{1cm}} DMF : N_2O_4$$

$$eq.\ 3b.)\quad DMF : N_2O_4 + CellOH \xrightarrow{\hspace{1cm}} solution$$

This solution would contain unchanged cellulose, $DMF:N_2O_4$ complex and uncomplexed DMF as the solvent. An additional possibility would be that dissolution may occur via some combination of the above mechanisms.

It follows that the mechanism of dissolution and thus the chemistry of the cellulose:N_2O_4:DMF interactions is very important since this will determine not only the composition of the spinning solution, spinning procedures, and fiber properties, but also the recovery and recycle process. Since some of the above mechanisms require complexation between the cellulose and N_2O_4 or between the DMF and N_2O_4, it appeared that during spinning the cellulosic fibers (rayon) might not require a chemical regenerating reaction but might be precipitated or coagulated by a nonsolvent system. This could be especially true for mechanism 3 since it requires that cellulose remain unchanged and be simply solvated by a $DMF:N_2O_4$ complex. However, if mechanism 1 were operative, then a transnitrosation reaction such as hydrolysis or alcoholysis would be necessary to obtain regenerated cellulose. One conceivable advantage of such a system is the possibility of control of molecular orientation during regeneration and spinning by control of the removal of the nitrite groups. Such increased control would result from the plasticity of the cellulose nitrite derivative permitting orientation and is responsible for the large variety of strong rayon fibers available from the viscose process, in which case the intermediate is cellulose xanthate. By contrast, if cellulose is merely precipitated or coagulated from a solvent the possibility of this control is severely limited.

As a further consequence, the nature of the ultimate spinning system is related to the dissolution mechanism by the actual composition of the cellulose/N_2O_4/DMF solution. If the spinning solution contained unreacted cellulose as in mechanism 3,

then a dry spinning system might be developed in which the cellu-
lose/N_2O_4/DMF solution is spun into an evacuated chamber to remove the solvents and precipitate the filaments. This would be
similar to the process for spinning cellulose diacetate from acetone solution. Advantages to this are the high speeds of spinning which are possible and the relative simplicity of the recovery system which would have only to recycle the DMF : N_2O_4
complex.

However, if cellulose had to be regenerated from a cellulose
derivative, in this case a nitrite ester, more complexity would
be involved in recovery and recycle. As previously explained,
regenerating cellulose during spinning from a true cellulose nitrite ester would require a transnitrosation reaction by some
agent to remove the nitrite and provide a hydrogen ion to cellulose. These requirements are met by protonic nucleophilic species
such as water, alcohol, or others. If the regenerant-coagulant
were water or alcohol, then the spent spin bath would contain
nitrous acid, HNO_2, or the alkyl nitrite, RONO, in addition to
the DMF and HNO_3 from the spinning solution, eq. 4. Any unreacted N_2O_4 in the spinning solution would make additional HNO_2
(or RONO) and HNO_3 in the spin bath, eq. 5.

4.) CellONO + H_2O (or ROH) $\rightarrow$ CellOH + HONO (or RONO)
     regeneration

5.) HOH (or ROH) + excess $N_2O_4 \rightarrow$ HONO (or RONO) + HNO_3

Total recovery and recycle, in this case, would involve splitting
the spin bath into at least 4 components and recycling DMF and
H_2O (or ROH) while changing RONO (or HONO) and HNO_3 into a form
from which N_2O_4 could be readily obtained. Thus, one of the
first questions to be answered was whether the spinning solution
contained cellulose nitrite.

Results and Discussion

This question was answered by chemical and spectral studies.
Because of the highly complex nature of the solution, simple
studies using reagents to measure nitrite composition were not
possible. The accuracy of any method using diazotizing reagents
or oxidation-reduction reactions to measure the concentration of
HNO_3 or CellONO would be hampered by the strong oxidizing ability
of the excess N_2O_4. Studies in which the spinning solution was
precipitated with water in a Waring Blender and the resultant
cellulosic material then analyzed for nitrogen and carboxyl residues showed that there was no increase in these over starting
pulp, but there was a slight loss in D.P. Thus any nitrite
derivative which was being formed during dissolution was not
stable to coagulation with water. This substantiated the claims

of earlier workers. Infrared studies were not pertinent since the nitrite (N=O) absorption occurs at ∼6.0μ which is in direct conflict with the amide carbonyl absorption of DMF so any spectral identification on an infrared basis would be considered tenuous.

Schweiger (1) had reported that a true cellulose nitrite ester could be isolated from the cellulose/N$_2$O$_4$/DMF solution by first stabilizing the solution with an acid scavenger such as pyridine and then precipitating the nitrite ester by water coagulation. These experiments were repeated. Pyridine was added to the 8/15/77 cellulose/N$_2$O$_4$/DMF solution at a level of 20% by weight of the solution and this was spread on a glass plate. The plate was then immersed in 20% pyridine/80% water at 5°C and an opaque film (cellulose always formed a clear film) was coagulated. This film was soluble in DMF, DMSO, CH$_3$CN and other organic solvents so it was obviously not cellulose.

The UV-VIS. spectrum in the 300 - 450 mμ region of a DMF solution of the re-dissolved film was nearly identical to that of isopropyl-, 1-pentyl- or cyclohexyl-nitrite which were prepared as model compounds by adding N$_2$O$_4$ to a DMF solution of the corresponding alcohol. When several drops of aqueous H$_2$SO$_4$ were added to the UV cell and the spectra were rescanned, all of the above, the model alkyl nitrites and the redissolved film, gave exactly the same spectrum. This new spectrum was identical to that of an acidified DMF solution of sodium nitrite indicating that all of these compounds, the alkyl nitrites and "cellulose nitrite", were releasing nitrous acid thus confirming that the new material did contain the nitrite moiety. Table I shows the UV-VIS. positions and intensities of various nitrite moieties used for comparison purposes in this structure proof. Figures 1 and 2 show the UV-VIS. spectra of DMF solutions of amyl nitrite, "cellulose nitrite", sodium nitrite and sodium nitrate before and after acidification. When the "cellulose nitrite" film was dried, brown gas evolved from the film and it then became insoluble in those solvents in which it had previously dissolved. The infrared spectrum of this insoluble dried film was identical to that of cellulose. This spectral study represents the first definitive proof that the material isolated by coagulation of the pyridine-stabilized cellulose/N$_2$O$_4$/DMF solution was indeed cellulose nitrite.

It was desirable to have chemical as well as spectral identification of the structure of this new material. In this regard, a method was developed for isolating a solid from the pyridine-stabilized solutions under anhydrous conditions. This was accomplished by the addition of diethyl ether to the pyridine stabilized cellulose/N$_2$O$_4$/DMF solutions causing precipitation of a solid. Addition of 1-pentyl alcohol to the DMF solution of this solid caused precipitation of a white fibrous material which was later identified as cellulose. Distillation

TABLE I

Positions and Intensities of Nitrite Ultraviolet
Absorption Peaks

Sample No.	Compound	Major Peaks max (mμ)	a	Minor Peaks or Shoulders max (mμ)	a
1.	Amyl Nitrite	370	67.2	385	32.2
		357	78.9	323	30.2
		345	67.2		
		333	46.8		
2.	Isopropyl Nitrite	373	55.4	385	31.2
		358	67.2	338	56.4
		348	67.2	328	(−)
3.	Cyclohexyl Nitrite	372	58.4	390	32.4
		360	57.8	348	44.5
				338	29.7
4.	"Cellulose Nitrite"	366		380	
		353		331	
		341		320	
5.	Sodium Nitrite	359	25.5		
6.	Sodium Nitrate	310	44		
7.	Sodium Nitrate with acid added	270	69		
8.	1, 3, 4 or 5 with acid added[b]	389	49.0		
		375	80.8		
		361	72.0		
		349	47.6		
		338	37.5		

a.) ϵ = molar extinction coefficient.

b.) The values for ϵ are from the experiment in which H_2SO_4 was added to the sodium nitrite solution.

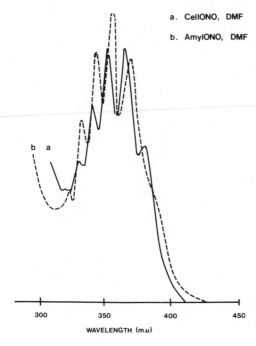

a. CellONO, DMF

b. AmylONO, DMF

WAVELENGTH (mμ)

Figure 1.   Uv–vis spectra of DMF solutions of amyl nitrite and cellulose nitrite

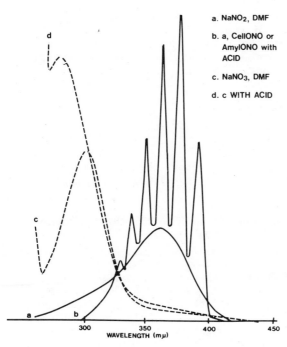

a. NaNO₂, DMF

b. a, CellONO or
AmylONO with
ACID

c. NaNO₃, DMF

d. c WITH ACID

WAVELENGTH (mμ)

Figure 2.   Uv–vis spectra of aqueous acidified DMF solutions of cellONO or Amyl-ONO as well as NaNO₂ and NaNO₃ before and after aqueous acidification

of the supernatant gave a 60% yield of 1-pentyl nitrite which was identical to commercial material in physical and spectral properties.

Therefore it was a certainty that the pyridine-modified cellulose/N_2O_4/DMF solution contained cellulose nitrite. It could be argued however, that the pyridine not only stabilized the cellulose nitrite but also catalyzed the formation of this species as pyridine is widely used in synthetic chemistry for catalysis in derivatizing alcohols (cellulose is an alcohol) with anhydrides (N_2O_4 is an anhydride). Spectroscopic studies on the cellulose/N_2O_4/DMF solution did not show that the nitrite was present, in fact UV-VIS. spectra of the solution are almost the same as those of nitrous acid. This is the result of the excess N_2O_4 which has basically the same UV-VIS. spectrum as nitrous acid probably because of the strongly absorbing nitrosyl (N=O) group. However, synthetic experiments indicated that model alcohols, such as cyclohexanol or isopropanol react immediately with DMF solutions of N_2O_4 at $5^{\circ}C$ in the absence of pyridine and since cellulose is an alcohol, it is expected to react similarly.

B. Recovery and Recycle of Raw Materials

Introduction

The entire area of research and development associated with the recovery and recycling of process chemicals is central to the technical and commercial success of spinning fibers from organic solvents. With the chemistry background presented earlier it appeared necessary to develop the recovery system based on processing the following:

    a.) Dimethylformamide (DMF)
    b.) Coagulant (either an alcohol or water)
    c.) Nitrogen Tetroxide (N_2O_4)

        1. When the coagulant is water, then nitrous acid (HNO_2) and nitric acid (HNO_3) are the precursors to N_2O_4 recovery.

        2. When the coagulant is an alcohol, then the alcohol nitrite (RONO) and HNO_3 are the precursors to N_2O_4 recovery.

An initial conceptualization of the recovery and recycle system is shown in Figure 3 which served as a preliminary sketch of the requirements of recovery and recycle for spinning rayon fibers from the cellulose/N_2O_4/DMF system. This figure shows that the crude spent spin bath, containing DMF, HNO_3, the coagu-

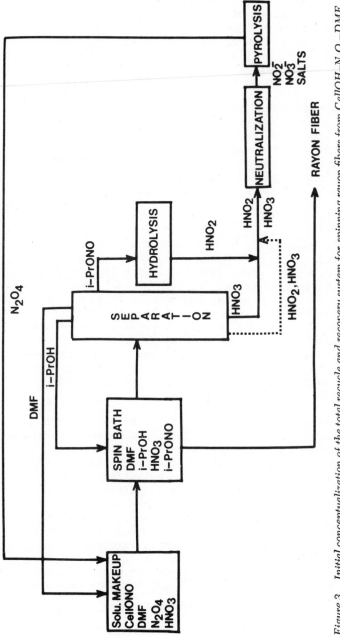

Figure 3.   Initial conceptualization of the total recycle and recovery system for spinning rayon fibers from CellOH–N₂O₄–DMF solutions

lent (either HOH or ROH) and the nitrosated coagulant (either
HONO or RONO) must be separated into fractions of pure DMF, pure
coagulant and the HNO_3-HNO_2 (RONO) fractions. The DMF must be
returned to the solution makeup, the coagulant to the spin bath
and the HNO_3-HNO_2 (RONO) portion of the spent bath must be combined in some way for conversion to N_2O_4. If this is the overall
plan, then two major areas of research would be necessary to
answer the following two questions: a.) which methods of separation would be most effective for the initial separation of the
crude spin bath and b.) how can the nitric and nitrous acid fractions resulting from cellulose dissolution and regeneration be
recombined to form N_2O_4. In addition, if the nitrous portion of
the original N_2O_4 is in the alkyl nitrite form, i.e. because of
an alcohol coagulant-regenerant, then a method for converting
HNO_3 and RONO to N_2O_4 would have to be found.

Results and Discussion

The recovery and recycle of N_2O_4 was studied first. An extensive literature survey on N_2O_4-HNO_3 chemistry uncovered several commercial processes which included a step for pyrolyzing a
metal nitrate salt to produce N_2O_4 in high yields. Most of these
processes involved decomposing calcium nitrate in a CO_2 atmosphere (2) or sodium nitrate in a CO atmosphere.(3) None of these
reports mentioned the recovery of N_2O_4 by pyrolyzing a nitrite
salt or a potassium salt as an industrial process.

Several methods were considered relative to the conversion
of the HNO_3 and HNO_2 portion of the spent spin bath to a salt
and recovery of this salt for pyrolytic processing. These
methods included:

a.) Precipitation of the salt - if the coagulant were
aqueous, then the spent spin bath containing HNO_2 and HNO_3 could
be neutralized to form salts. Following this, addition of a
suitable non-solvent such as alcohol could precipitate the salts.
This presented unique problems since the solubility of KNO_3 is
relatively small (1.5%) in DMF, but that of $Ca(NO_3)_2$ or $NaNO_3$ is
$\sim$ 15-20%. If the spin bath contained alcohol as the coagulant
then a hydrolysis step, to convert RONO to HONO, would precede
the acid neutralization. However, in such a non-aqueous bath,
the possibilities of using other water immiscible precipitants
such as methylene chloride or ether were recognized.

When studies were done, effective methods of precipitating
KNO_3 and $NaNO_3$ but not $Ca(NO_3)_2$ were found as can be seen in
Table II.

TABLE II

Solubilities of Nitrate Salts in DMF/Precipitating Solvent Mixtures[a,b]

Precipitating[c] Solvent	%Precipitating Solvent[d]/ Salt Left Dissolved in Mixture[e]			
	KNO_3	$NaNO_3$	$Ca(NO_3)_2$	$Pb(NO_3)_2$
None (Exp.)	0/1.1	0/14.2	0/18.0	0/130
None (Lit.)(4)	0/1.5	0/15.4	N. A.	N. A.
N_2O_4	15.7/ 0.5	47.5/ 1	50/18.0	49/3.3
MeOH	63.4/1.1	41.5/12.2	53.5/18.0	75/42.4
EtOH	N. D.	59.7/4.3	N. D.	N. D.
i-PrOH	61.5/ 0.5	60.3/ 1	56.5/18.0	64/ 1
Ether	59.4/ 0.5	39.7/ 1	61.3/18.0	N. D.
CH_2Cl_2	73/ 0.5	67/1.2	N. D.	N. D.

a) N. A. = not available
b) N. D. = no data obtained
c) N_2O_4 = dinitrogen tetroxide, MeOH = methyl alcohol, EtOH = ethyl alcohol, i-PrOH = isopropyl alcohol, CH_2Cl_2 = methylene chloride, ether = diethylether
d) % Precipitating solvent based on total mixture
e) g. salt/100g DMF

b.) Distillation of all liquid to leave a salt residue – if the coagulant were aqueous, then neutralization of the acids followed by distillation to a solid salt residue (the potential for detonation at this point due to the nitrate–DMF mixture was recognized early in our work) would appear to be viable. If the coagulant were alcohol, then as stated above, the hydrolysis step would precede the neutralization step. Alternately, the HNO_3 portion of the spin bath could be neutralized prior to distillation, the spin bath distilled to a salt residue and the alcohol nitrite portion recovered from the distillation, hydrolyzed, neutralized and then the nitrite salt recovered after liquid stripping. Literature available on the recovery of DMF from aqueous systems indicated that acceptable yields could be obtained by distillation.(5)

The distinct difference between schemes (a) and (b) was that in scheme (a), the DMF would not require distillation. Hopefully this would have led to lower economics for scheme (a) than scheme (b).

c.) Countercurrent extraction of the aqueous–DMF spent spin bath – this would only apply for aqueous spin baths. Commercial methods for countercurrent extraction of DMF from aqueous–DMF solutions were described in detail in the literature.(6)   A variety of extractants including chlorocarbons and hydrocarbons were evaluated. The conclusion of this literature study was that methylene chloride was the most efficient extractant and that the process was economically feasible when the aqueous–DMF contained less than 10% DMF. Results in this laboratory did not show pro- mise probably because the aqueous DMF spent spin baths contained acid or salt components which complicated extraction.

Recovery of the coagulant portion of the spin bath, i.e. water or an alcohol would be, in a sinse, a limiting factor since, depending on the composition of the spin bath, very large volumes would be involved in the recovery as will now be ex- plained. Although some early spinning work was done using 8/25/67 cellulose/N_2O_4/DMF solutions, a decision was made, based on comparative studies between 8/25/67 and 8/15/77 spinning solu- tions, to make the detailed process evaluation on an 8/15/77 cell- ulose/N_2O_4/DMF solution. This requires that 1.88 lbs. of N_2O_4 and 9.63 lbs. of DMF be recycled per 1.00 lb. of processed fiber in addition to the coagulant portion of the spin bath. For ex- ample, if the spin bath were at equilibrium and were 58/31/5/6, isopropyl alcohol (i-PrONO/DMF/HNO_3/isopropyl nitrite (i-OrONO) then a total of $\sim$ 28.3 lbs. (9.63 lbs. DMF and 18.64 lbs. of i-PrONO) of liquid would have to be recycled per 1.00 lb. of fiber processed.

If the spin bath were aqueous–DMF at a level of 20% H_2O then a minimum of 12.0 lbs. of liquid would be recycled per 1.00 lb. of processed fiber. These represent examples of how the spin bath composition sets a lower limit on the amount of chemi- cals which must be processed. Therefore the economics of a com- mercial process depend greatly on the spin bath composition.

Initial spinning experiments indicated that alcohol–DMF spin baths produced better fibers than aqueous–DMF spin baths. Pure methyl-, ethyl- or isopropyl-alcohol as coagulant–regenerants produced fibers with approximately equivalent physical properties. Since methyl nitrite is a toxic gas and ethyl nitrite has a very low boiling point, they are difficult to handle at an experi- mental level. However, isopropyl nitrite (i-PrONO) is reason- ably easy to handle and therefore i-PrOH was chosen as the co- agulant–regenerant for detailed recovery studies. The physical properties of fibers coagulated in alcohol–DMF systems decreased as the level of DMF increased, the best fibers having been spun from pure alcohol primary spin baths. However, it was apparent that, for economic reasons, some level of alcohol in DMF would have to be chosen and the fiber properties maximized for this chosen system. An i-PrOH/DMF ratio of 2/1 was chosen and thus when recovery studies were begun, synthetic spent spin baths of

TABLE III

Recovery of Process Chemicals by Vacuum Distillation of the Crude Spin Bath

Fraction No.	Conditions bp. °C/mm. Hg	Total Wt. g.	DMF g.	i-PrOH	i-PrONO g.	HNO_3 g.	H_2O g.
1	r.t./10	95.9	–	18.9	76.9	0.0	–
2	25–32/1	88.7	3.4	82.9	0.0	0.0	2.4
3	32–46/1	76.4	4.9	62.5	0.0	0.0	5.1
4	bottoms	120.6	67.1	0.0	0.0	39.8	11.0
totals		381.6	75.4	164.3	76.9	39.8	18.5
starting spin bath		408.7	78.0	165.2	80.5	41.1	23.4
% accounted for		93.4	96.7	99.5	95.5	96.8	79.1

a.) This 76.9g. represents 52.7g. i-PrOH

b.) This 67.1g. of DMF is composed of DMF which is available for recycle by simply distilling the liquid (28.6g.) and DMF which is bound to the HNO_3 and can be recovered if the complex is broken by neutralization.

2/1 i-PrOH/DMF in which N_2O_4 was added to bring the HNO_3 level
near 3, 5 and 10% were evaluated. These solutions were
vacuum distilled without neutralization to establish exactly how
such an acidic solution would behave in a recovery process.
Table III shows a synopsis of the distillation of a spin bath at
the 10% acid level.

The i-PrONO and i-PrOH distilled together as the pressure
was reduced. The next fraction contained i-PrOH/DMF and quite
unexpectedly, no HNO_3 was found in this fraction. In fact, pure
DMF was obtained as distillate in the next fraction until the
bottom residue contained nearly 1.0 mole HNO_3/1.0 mole DMF.
This residue was fractionally distilled to give a clear, color-
less, oily liquid, bp. 94-95°C/8 mm Hg. The liquid had a density
of 1.21 g/ml and contained 47.07% HNO_3 (titration with standard
caustic) and 48.98% DMF (gas chromatographic analysis). The
theoretical values are 1.14 g/ml as the density and 46.32% HNO_3,
53.68% DMF. Thus, this new material represents a 1:1 molar mixture of HNO_3 and DMF in which there is obviously some interaction
between the two molecular species, i.e. complexation to some extent. This interaction is also suggested by nuclear magnetic
resonance (NMR) studies which showed a downfield shift of the
DMF-aldehyde and N-methyl protons in the mixture as compared to
these protons in pure DMF. As these spectra were taken in the
neat liquids, no solvent effects could interfere with the results. This downfield shift was also present when the spectra
were taken in $CDCl_3$/TMS. A tabulation of these shifts are shown
in Table IV.

TABLE IV

NMR Shifts in DMF and DMF/HNO_3 Complex

Sample	δ (ppm) [a.]		
DMF-HNO_3	3.09	3.24	8.28
DMF	2.80	2.98	8.03
	0.29	0.26	0.25

a.) Chemical shifts in δ (ppm) downfield from the internal standard TMS.

The DMF-HNO_3 binary mixture is intriguing since it provides
a method of separating most of the DMF from the HNO_3 even
though the boiling point of DMF (154°C) is much greater than that
of HNO_3 (83°C). The downfield chemical shift of the aldehyde and
N-methyl resonances indicate a decrease in electron density at

the amide carbon atom which is consistent with a weak complexing of the amide and HNO_3. Attempts to dissolve cellulose in this binary mixture were unsuccessful.

Thus vacuum distillation of a synthetic spin bath at concentrations expected in a commercial process gave separation of all components except that portion of the DMF which was involved in a DMF/HNO_3 binary mixture. This DMF could be released by addition of a base and various bases were studied for this purpose but those derived from calcium were chosen for several reasons as explained below. Calcium carbonate ($CaCO_3$), calcium hydroxide ($Ca(OH)_2$) or calcium oxide (CaO) were used to release the DMF from this DMF/HNO_3 mixture. The resultant clear thick solution was vacuum distilled to recover the DMF leaving a calcium nitrate ($Ca(NO_3)_2$) residue. This solid residue gave N_2O_4 when pyrolyzed. Thus, this sequence established technically that it was possible to fractionate the spin bath, recover i-PrOH, i-PrONO and DMF from HNO_3 and obtain N_2O_4 from the HNO_3 portion of the bath.

The choice of calcium bases for neutralization results from a study which was done on the stability of DMF under acidic and basic conditions. The data clearly showed that DMF is reasonably stable in the presence of acid as long as water is excluded but that dimethylamine (DMA) is formed by hydrolysis when water is present. By contrast, DMF decomposed rapidly when NaOH or KOH was added with no additional H_2O. When $Ca(OH)_2$, CaO or $CaCO_3$ was added to DMF, even with additional H_2O, no measurable amount of DMA was formed even on prolonged standing at room temperature. In addition, the recovery of N_2O_4 from $Ca(NO_3)_2$ by pyrolysis is a commercial process related to the phosphoric acid industry and literature references were found to indicate that good yields of N_2O_4 from pyrolysis of $Ca(NO_3)_2$ were customary. [2]

Following an extensive literature search, a project was established to study the production of N_2O_4 via pyrolysis of various metal nitrates and nitrites. A muffle furnace was modified to hold a pyrolysis tube containing the salt under study. A stainless steel pyrolysis chamber was built so that different sweep gases could be used to create the desired atmosphere in the pyrolysis tube and to sweep the product N_2O_4 gases into a capturing solution of 1-pentyl alcohol in DMF. The amount of N_2O_4 could then be determined by UV–VIS. analysis of the amount of 1-pentylnitrite formed in the capturing solution. References in the literature indicated good N_2O_4 yields could be obtained from $Ca(NO_3)_2$ at 600°C in a CO_2 atmosphere [2] or $NaNO_3$ in a CO atmosphere [3] but no data were available on the pyrolysis of KNO_3 or nitrite salts. In addition to obtaining the N_2O_4 yield, it was necessary to measure the content of unpyrolyzed nitrate and nitrite in the pyrolysis salt residue to determine the selectivity of the process. Methods for these determinations were developed.

The expected salt mixture which would be obtained from the recovery process would be an equimolar mixture of calcium nitrate/calcium nitrite. Nitrite salts decompose to produce one mole of NO and one mole of NO_2 per mole of nitrite salt thus additional oxidation is required to obtain N_2O_4 from this system.

Equations 6 and 7 are examples of the chemical changes accompanying the release of N_2O_4 during the decomposition of $Ca(NO_3)_2$. Although the byproducts vary with the different sweep gases and salts, the decomposition of calcium nitrate, $Ca(NO_3)_2$ and calcium nitrite, $Ca(NO_2)_2$ in the presence of CO_2 should serve as adequate examples.

6. $Ca(NO_3)_2 + CO_2 \longrightarrow CaCO_3 + 2NO_2 + 1/2\ O_2$

7. $Ca(NO_2)_2 + CO_2 \longrightarrow CaCO_3 + NO_2 + NO$

When a mixture of nitrate and nitrite is pyrolyzed, 1/2 mole of oxygen is produced, which is the required amount to convert the NO from nitrite decomposition to N_2O_4. However, in the laboratory experiments, oxygen was mixed with the effluent gases in oxidation chambers which were placed before the capturing chambers for all pyrolysis runs which included nitrite. This assured the best possible yields. Yields of N_2O_4 resulting from pyrolysis experiments are shown in Table V. The results in this table show that 93% N_2O_4 yields can be obtained from the pyrolysis of a 1:1 $(Ca(NO_3)_2/Ca(NO_2)_2$ mixture at $800^{\circ}C$ in a CO_2 atmosphere. If the HONO or $Ca(NO_2)_2$ could be oxidized to HNO_3 or $Ca(NO_3)_2$, then the N_2O_4 recovery would involve the pyrolysis of $Ca(NO_3)_2$ which routinely gave 98% yields of N_2O_4 in CO_2 or N_2 at $800^{\circ}C$. This part of the investigation established a route to N_2O_4 recovery, i.e. by pyrolysis of a $Ca(NO_3)_2/Ca(NO_2)_2$ salt mixture.

Since the N_2O_4 which is produced by this pyrolysis is a corrosive, oxidizing gas, a specific study was made to determine the requirements for construction materials for the pyrolysis furnace. Conferences with faculty members at the Engineering School, Rutgers University, New Brunswick, N.J. resulted in the suggestion that a ceramic material would be required for the inside of the furnace. The material of preference was Al_2O_3 which is frequently used for furnace linings. Prototype laboratory pyrolysis tubes of Al_2O_3 were designed and obtained from Duramic Products, Inc. Preliminary pyrolysis data from a limited number of experiments using pure calcium nitrate indicate that the yields of N_2O_4 are similar to those obtained when stainless steel tubes were used.

One of the important remaining steps was conversion of isopropylnitrite into the calcium nitrite salt for pyrolysis. Hydrolysis of i-PrONO to HNO_2 (nitrous acid) and neutralization of

TABLE V

The Effect of Temperature and Sweep Gas on N_2O_4 Yields from the
Pyrolysis of Nitrate Salts or 1:1 Nitrate/Nitrite Salt Mixtures[a,b]

Salt/Temp. °C/Time (min.)	Sweep Gas			
	Carbon Dioxide	Carbon Monoxide[f]	Nitrogen	Air
NaNO₃/800/90[c]	75.3/86.5	35.0/58.0	9.6/63.9	-/-
KNO₃/800/90[c]	67.7/84.8	11.9/36.6	6.0/69.2	-/-
Ca(NO₃)₂/600/120	93.3/100.1	-/-	38.2/95.2	-/-
Ca(NO₃)₂/800/90	97.8/97.8	-/-	98.1/98.1	-/-
NaNO₃:NaNO₂/1000/90[e]	76.8/84.6	53.1/54.1	20.1/31.7	-/-
KNO₂:KNO₂/1000/90[d,e]	61.9/68.8	28.8/33.9	28.8/38.5	13.9/41.0
Ca(NO₃)₂:Ca(NO₂)₂/600/120[e,g]	57.8/-	-/-	34.0/-	26.0/-
Ca(NO₃)₂:Ca(NO₂)₂/800/90[e,g]	92.8/-	-/-	78.8/-	84.4/-

a) These results are averages of 2-4 pyrolysis experiments. If the salts were not thoroughly dried, lower N_2O_4 yields resulted. The sweep gas flow rates were 100 ml./min. Stainless steel (310) pyrolysis chambers were used.

b) Results are expressed as % yield of N_2O_4/total % of the starting salt which is accounted for by the sum of the N_2O_4 yield plus analysis of the pyrolysis residue.

c) Pyrolysis at 600°C resulted in very low N_2O_4 yields.

d) Pyrolysis at 800°C resulted in very low N_2O_4 yields.

e) The pyrolysis of a nitrite salt produces a mole of NO per mole of NO_2. Therefore O_2 was mixed at 50 ml./min. with the pyrolysis effluent to oxidize any NO to N_2O_4 for capture.

f) Carbon monoxide can reduce NO_2 to NO by the equation $CO + NO_2 \rightarrow NO + CO_2$ therefore O_2 was mixed at 50 ml./min. with these pyrolysis effluent streams.

g) When these measurements were made, no method of residue analysis was available.

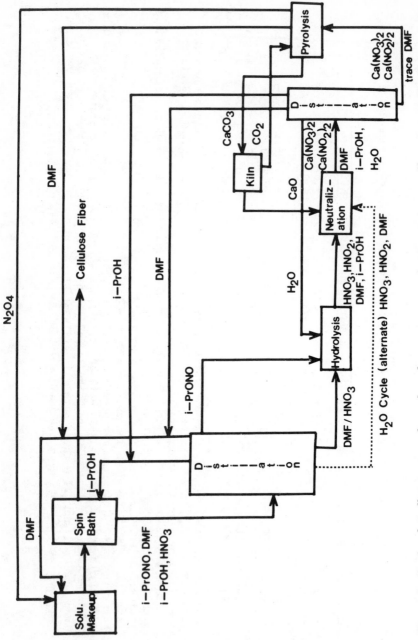

Figure 4. Final technically proven total recycle and recovery system for spinning rayon fibers from CellOH–N₂O₄–DMF solutions

this HONO appeared reasonable. Although this approach was followed, the problem of efficient conversion of i-PrONO to HONO
remained one of the weak points in the recovery scheme. After
detailed study, it appeared that the choice medium for the hydrolysis was the HNO_3/DMF obtained as the bottoms fraction from
distillation of the crude spin bath. This was expected since
hydrolysis is acid catalyzed. In addition, the DMF caused the
i-PrONO to be miscible with the H_2O added for hydrolysis whereas
normally i-PrONO and H_2O are not miscible.

The progress in development of this aspect of the recovery
cycle was impeded by one particular factor. A method for monitoring either the concentration of i-PrONO or HNO_2 in the
DMF/HNO_3/H_2O hydrolysis solutions was not known. The UV-VIS.
spectra of i-PrONO and HNO_2 are overlapping so that this was not
a useful method. A method for gas chromatographic analysis of
nitrites, (isopropyl, amyl, isoamyl or cyclohexyl) was not found
since the nitrite apparently decomposes during chromatography.
In fact, under all of the conditions attempted, a peak for the
parent alcohol was observed when the nitrite was injected. In
addition, several peaks, attributable to nitrogen oxide gases
were present in these chromatograms. The temperature of the
injection port and the column as well as the gas flow rate were
changed as was the column composition but no suitable conditions
were found under which pure nitrite did not fully decompose to
its parent alcohol. One method for analysis may be NMR spectroscopy but this was not readily available. Thus, the amount of hydrolysis was crudely measured by aqueous titration with standard
caustic.

Mixtures of nitrites, isopropyl as well as model alkyl nitrites, were studied in DMF and H_2O with or without acid. A
mixture of i-PrONO/H_2O/DMF was slurried for 15 minutes at room
temperature and then neutralized with $Ca(OH)_2$ or CaO. It was
not efficient to use $CaCO_3$. This new mixture was vacuum distilled to give a powdery residue which gave N_2O_4 on pyrolysis.
Additional experimentation indicated that it was possible to recover unhydrolyzed nitrite by careful fractional distillation of
the neutralized isopropyl nitrite hydrolysis mixture. A total
recovery of HNO_2 (by titration) plus the recovered unchanged
i-PrONO (by distillation) was 94%, suggesting the possible use
of multiple hydrolysis units.

The entire recovery and recycle scheme based on this work is
shown in Figure 4. This is essentially similar to that proposed
in Figure 3 except that all of the loops have been closed. A recovery and recycle scheme such as this should not release any
effluent to the environment because it is totally cyclical with
all steps included in closed loops.

Conclusions

It has been definitively proven that the material which is
recoverable by coagulation of a pyridine stabilized cellulose/

N_2O_4/DMF solution is cellulose nitrite. However, although it seems reasonable that cellulose nitrite is formed in the absence of pyridine this factor does not limit the utility of the proposed recycle and recovery scheme because that scheme is dictated by the spinning system which is required to make the most commercially acceptable rayon fibers. In this case, the spinning system of choice was based on isopropyl alcohol and this alcohol immediately forms isopropyl nitrite when it contacts the cellulose/N_2O_4/DMF solution irregardless of whether the cellulose is complexed or derivatized.

All chemical steps required to recover and recycle the DMF, coagulant and N_2O_4 from the spent spin bath have been delineated and proven to be feasible from a technical standpoint.

Abstract

The dissolution of cellulose in DMF/N_2O_4 is discussed in terms of three possible mechanisms, one involving derivatization of the cellulose and two describing dissolution in terms of complex formation. Chemical and spectral studies on the pyridine stabilized cellulose/N_2O_4/DMF solution strongly support the mechanism of derivatization of the cellulose by N_2O_4.

A total recovery and recycle scheme has been proposed and proven to be technically feasible. This scheme involves separation of the crude spent spin bath, which contains DMF, N_2O_4, i-PrOH and i-PrONO, into four major fractions. These are a.) pure i-PrONO, b.) pure i-PrOH, c.) pure DMF and d.) a 1:1 molar DMF/HNO_3 complex. The DMF/HNO_3 complex is used to catalyze hydrolysis of the i-PrONO to i-PrOH and HNO_2. This HNO_3/HNO_2 combination is neutralized with CaO and the resultant $Ca(NO_3)_2$/$Ca(NO_2)_2$ is pyrolyzed in a CO_2 atmosphere to recover N_2O_4 for recycle. Calcium carbonate resulting from the pyrolysis of $Ca(NO_3)_3$/$Ca(NO_2)_2$ in CO_2 is converted to CaO and CO_2 to continue the cyclical process.

Literature Cited

1. Schweiger, R.G., U.S. Patent No. 3,702,843 (1972).
2. Delassus, Marcel; Copin, Robert; Hofman, Theophile; Sinn, Robert, S. African Patent No. 68 0155809 (Aug. 1968). CA 70 P 89317z.
3. Industrie-Werke Karlsruhe Akt. Ges. German Patent No. 1,014,086 August 22, 1957 CA 53P11412C.
4. Monograph-DMF: A Review of Catalytic Effects and Synthetic Applications, E.I. duPont de Nemours and Co., Inc., page 3.
5. Monograph-DMF: Recovery and Purification, E.I. duPont de Nemours and Co., Inc., page 10.
6. Monograph-DMF: Recovery and Purification, E.I. duPont de Nemours and Co., Inc., page 12.

Production of Rayon Fibers from Solutions of Cellulose in (CH₂O)x-DMSO

R. B. HAMMER, A. F. TURBAK, R. E. DAVIES, N. A. PORTNOY

ITT Rayonier, Inc., Eastern Research Div., Whippany, N.J. 07981

A wide range of cellulosic pulps in various forms were found to dissolve readily at elevated temperatures in a combination of $(CH_2O)x/DMSO$. The concentration of pulp dissolved was a direct function of the degree of polymerization. In general, 6/6/88 cellulose/$(CH_2O)x$/DMSO solutions were prepared using pulps of 400 to 600 D.P. The solutions were microscopically free of gels and undissolved cellulosic fibers.

This dissolution process was found to be very specific to the combination of $(CH_2O)x$/DMSO. Other analogous aldehydes and polar organic solvents failed to afford cellulose solutions.

Cellulosic articles such as fibers and films are easily regenerated from the cellulosic solution in the presence of aqueous solutions having a pH greater than seven, of water soluble nucleophilic compounds such as ammonia, ammonium salts, and saturated amines. Salts of sulfur compounds in which the sulfur has a valence of less than six may also be used. Fibers of high wet modulus, intermediate tenacity and low elongation were readily produced from regenerating systems such as aqueous ammonia, ammonium carbonate, tetramethyl ammonium hydroxide, methyl amine, diethyl amine, trimethyl amine, sodium sulfide, sodium sulfite and sodium thiosulfate.

Fibers have been spun with conditioned and wet tenacities as high as 2.7 and 1.5 g/d respectively with wet modulus of as high as 1 g/d and solubility in 6.5% NaOH as low as 3-15%.

Introduction

Today, rayon is almost universally produced by the viscose process. However, the high investment costs and potential mill effluent pollution problems associated with viscose rayon plants makes this process increasingly less competitive from both an economic and environmental standpoint. There are other processes for producing regenerated cellulosic products including regeneration from cellulose nitrate which is very hazardous or from cuprammonium hydroxide. However, the production of cellulosic articles from these processes is minuscule compared to that from the viscose process.

A number of highly polar, aprotic organic solvents for cellulose have been disclosed in the literature. Two solvents which have received frequent mention are dimethylformamide (DMF)(1), (4)-(8) and dimethyl sulfoxide,(1),(2),(3) each in combination with one or more additional compounds such as N_2O_4(1)-(7), SO_2(8)-(9) or an amine.(10) More recently, DMSO-paraformaldehyde has been reported as a solvent for cellulose.(11)

While there has been much discussion of these and other solvent systems for cellulose, the literature contains little information concerning the regeneration of fibers, films or other regenerated cellulosic articles from such solvent systems. There are almost no data in the literature, for example, on the properties of fibers spun from an organic solvent system. In so far as is known, no economically acceptable commercial solvent-based processes have as yet been disclosed for producing fibers or films with acceptable properties.

Experimental

Experiments were designed to determine the influence of the form of the pulp and the degree of polymerization prior to dissolution. Although the majority of the solutions were prepared using Abbe' cut material, i.e. a highly comminuted, defibered pulp, the dissolution process is not limited with respect to the degree of polymerization or the pulp form.

All solution compositions are given as weight percents in the order, pulp/(CH_2O)x/DMSO eg. 6/6/88 represents 6% cellulose, 6% paraformaldehyde and 88% solvent. A typical example of the solution preparation procedure is described below.

Silvanier-J, a prehydrolyzed kraft pulp of 1050 D.P., after converting to alkali cellulose by methods well known in the rayon industry, was alkaline aged to a D.P. level of 450, neutralized, washed, dried, then either fluffed, diced or defibered.

A 6/6/88 cellulose/(CH_2O)x/DMSO solution was prepared by charging 120 parts of alkali aged prehydrolyzed kraft pulp (D.P. 450), 120 parts of powdered paraformaldehyde and 1760 parts of DMSO into a two-liter four neck resin reaction flask equipped with a stainless-steel mechanical stirrer and thermometer. The resulting slurry was stirred and heated to 120°C over a period of one hour. Although dissolution is almost complete at 120°C after about one hour, the heating and stirring were continued for 1) an additional hour at 120°C or 2) the length of time required to remove excess formaldehyde.

The cellulose/(CH_2O)x/DMSO solutions described herein were of 6/6/88 composition. By employing low D.P. pulps ie in the range of 200-400 D.P. it is possible to prepare DMSO/(CH_2O)x solutions containing 10-12% cellulose. Solutions containing 8-10% cellulose can be prepared by using pulps with D.P. levels between 400 and 600 but the resulting viscosities exceed 30,000 cps at ambient temperature.

All the cellulose/(CH_2O)x/DMSO solutions were observed to be microscopically free of gels and unreacted fibers. The solutions were deaerated prior to spinning and viscosities measured by a Brookfield Viscometer. They were in the range of from 8,000-20,000 cps. at ambient temperatures for pulps with 400-600 D.P. levels. The solutions were filtered through a 90 mm diameter nylon, in-line filter during spinning.

Several types of spinnerettes have been used successfully to spin fibers from this solvent system. For example, gold-platinum typical for the viscose process, stainless steel typical for cellulose acetate spinning, and glass. Glass spinnerettes are preferred since they are inexpensive.

The majority of the spinning trials were performed using a bench-scale vertical spinning unit. The solutions were spun into the appropriate primary regeneration bath and the resulting fiber tow passed vertically to a primary glass godet then through a secondary bath to a secondary glass godet whose speed could be altered to produce the desired stretch conditions.

Spinning speeds generally ranged between 10 and 70 meters/minute but no attempt was made to optimize this condition or the resulting fiber physical properties.

The cellulose/(CH_2O)x/DMSO solutions regenerate rapidly in aqueous solutions of nucleophilic species which act as formaldehyde scavengers. This class of compounds includes ammonia, ammonium salts, saturated amines and salts of sulfur compounds.

All fibers were processed as staple by treatment with 60-70°C water, a 0.3% aqueous solution of a finishing agent at 50°C, then centrifuged and oven dried at 100-110°C. The fibers were tested for physical properties according to the American Society of Testing and Materials standards D-1577-73 and D-2101-72.

Results and Discussion

The solvent combination of (CH_2O)x/DMSO will dissolve a wide range of cellulosic pulps in either fluffed, diced or Abbe' cut form. The percent cellulose in the resulting solutions is dependent upon the pulp D.P. In general, 6/6/88 cellulose/(CH_2O)x/DMSO spinnable solutions can be prepared from pulps within the 400-600 D.P. range. The pulps may be either sulfate or sulfite grades and include many of the pulps that are typically employed in the viscose process. Ground wood in the form of newsprint did not dissolve in the paraformaldehyde/DMSO combination under the conditions employed.

The dissolution of cellulose in the (CH_2O)x/DMSO solvent system is believed to result from the formation of a hemiacetal of cellulose ie methylolcellulose.(11)

The dissolution of cellulose was found to be very specific
to the combination of DMSO and paraformaldehyde. Organic solvents
analogous to DMSO such as sulfolane, sulfolene or DMF did not
cause dissolution in the presence of paraformaldehyde. Likewise,
aldehydes similar to formaldehyde or paraformaldehyde, for example chloral, acetaldehyde, benzaldehyde, and compounds such as
trioxane, methyl formcel and butyl formcel in the presence of
acid catalysts did not yield cellulose solutions in combination
with DMSO.

Cellulose/(CH_2O)x/DMSO solutions can be rapidly coagulated
by employing nucleophilic species which act as formaldehyde-
scavenging agents. In general, these regenerants are aqueous
solutions of water soluble nucleophilic compounds which have a
pH greater than seven. Examples of useful regenerants include
ammonia, ammonium salts, saturated amines and salts of sulfur compounds in which the sulfur has a valence of less than six.

In the case of the nitrogenous compounds, the coagulant is
actually ammonia or an amine, the source of which may be, in
addition to ammonia or the amine itself, an ammonium salt, or, in
some instances, a basic amine salt. Under the alkaline conditions
of the regeneration solution, ammonium or amine salts will hydrolyze to liberate the free base ie ammonia or the amine, respectively.

A particularly useful nitrogenous compound is ammonium
hydroxide. If the regeneration bath is aqueous ammonium hydroxide, then after regeneration of a fibrous tow, the spent bath
would contain water, DMSO, ammonia and hexamethylenetetramine.
The latter compound would result from the reaction of ammonia and
formaldehyde. Other nitrogenous compounds which possess the
requisite nucleophilic, solubility and pH properties are salts of
ammonia and a weak acid such as ammonium acetate, ammonium sulfide, ammonium carbonate and ammonium bisulfite. Amines which are
useful are, in general, saturated aliphatic, cycloaliphatic and
alicyclic amines. Aromatic amines and amines of more than six
carbon atoms are normally insoluble or of borderline solubility
in water.

Particularly effective sulfur compounds are sodium sulfide,
sodium sulfite and sodium thiosulfate. The sulfates, in which
sulfur has a valence of six are not useful. An amount of the
nucleophilic compound as little as 0.25% by weight of the regeneration solution has been found effective for regeneration of the
cellulose. The maximum concentration is limited only by the solubility of the compounds in water. Normally the concentration will
range from 3-15%.

Variations in the composition of the regeneration or spin
bath did not alter the fiber cross-sections. Both the nitrogenous and sulfur containing regenerants afford rayon fibers with
circular shapes. Representative Shirley cross-sections of fibers
spun from cellulose/(CH_2O)x/DMSO solutions are illustrated in

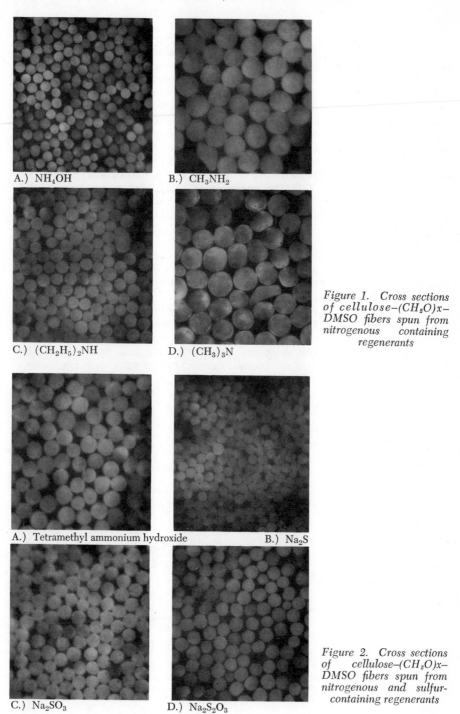

A.) NH$_4$OH

B.) CH$_3$NH$_2$

C.) (CH$_2$H$_5$)$_2$NH

D.) (CH$_3$)$_3$N

Figure 1. Cross sections of cellulose–(CH$_2$O)x– DMSO fibers spun from nitrogenous containing regenerants

A.) Tetramethyl ammonium hydroxide

B.) Na$_2$S

C.) Na$_2$SO$_3$

D.) Na$_2$S$_2$O$_3$

Figure 2. Cross sections of cellulose–(CH$_2$O)x– DMSO fibers spun from nitrogenous and sulfur-containing regenerants

Figures 1 and 2.

The regenerated cellulosic fibers produced from some of these nitrogenous or sulfur containing compounds are fully comparable in properties to cellulosic fibers produced by the viscose process. They are particularly outstanding in having a very low "$S_{6.5}$". The $S_{6.5}$ values are measurements of regenerated cellulosic fiber solubility in 6.5% sodium hydroxide at 20°C. This is a useful test for determining the potential resistance of such fibers or resultant fabrics to alkaline treatment such as alkaline laundering or mercerization. Accordingly, regular viscose rayon which cannot be mercerized and is not resistant to alkaline washing (unless cross-linked), has a relatively high $S_{6.5}$ of from 25-35%. On the other hand, the high performance and polynosic rayons have superior resistance to caustic soda as evidenced by $S_{6.5}$ values of from 5-15%. Cellulosic fibers spun from cellulose/$(CH_2O)x$/DMSO solutions have $S_{6.5}$ values in the range of from 3-15%.

It should be noted here that in this spinning system it is readily possible to obtain fibers with high wet modulus without the use of zinc or other additives which are required in a viscose spinning operation.

In Figure 3 stress-strain curves [conditioned (c) and wet (w)] for a sodium sulfide spun fiber are shown for comparison with regular rayon and high wet modulus rayon.

Some typical fiber physical property data are shown in Table I employing a variety of primary regeneration baths for the pro-

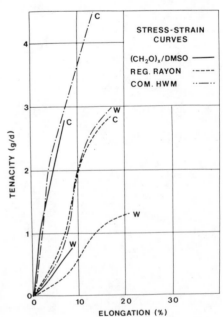

STRESS-STRAIN
CURVES

$(CH_2O)_x$/DMSO ———
REG. RAYON ------
COM. HWM ···—··

TENACITY (g/d)

ELONGATION (%)

Figure 3. Stress-strain curves of regular rayon, commercial high wet modulus rayon and rayon produced by regenerating cellulose–(CH_2O) x–DMSO solutions from aqueous sodium sulfide

TABLE I

PROPERTIES OF FIBERS PREPARED FROM A 6/6/88 CELLULOSE/$(CH_2O)x$/DMSO SOLUTION

Coagulant	Denier	Tenacity, g/d		Elong., %		Wet Modulus g/d	% $S_{6.5}$
		Cond.	Wet	Cond.	Wet		
2.9% NH_4OH	4.1	2.2	0.69	6.1	10.3	0.30	11
15% Ammonium Carbonate	0.86	2.7	1.21	4.1	10.4	0.50	–
4% Na_2S	0.60	2.58	1.32	5.0	7.4	0.85	9.0
20% Tetramethyl ammonium hydroxide	2.79	2.05	0.82	17	18	0.20	–
10% Na_2SO_3	1.94	2.46	0.86	6.8	8.0	0.47	–
10% CH_3NH_2	4.2	1.64	0.71	4.4	12.5	0.25	–
10% $(CH_2H_5)_2NH$	1.7	1.44	0.77	6.1	9.9	0.34	14.1
10% $Na_2S_2O_3$	1.82	2.43	1.18	5.5	7.1	0.65	6.2
12.5% $(CH_3)_3N$	4.8	1.01	0.67	13.2	18.1	0.17	–

TABLE II

FIBER PHYSICAL PROPERTIES

Staple Type	Tenacity, g/d		Elongation, %		Wet Modulus g/d	$S_{6.5}$, %
	Cond.	Wet	Cond.	Wet		
Regular Rayon	1.5-2.8	1.0-1.8	14-25	18-35	0.18-0.28	20-35
Intermediate Wet Modulus Rayon	3.5-5.0	2.5-3.5	12-19	18-24	0.45-0.60	15-20
High Wet Modulus Rayon	3.5-8.0	2.5-6.0	6-14	9-18	0.7-3.0	5-10
Prepared from 6/6/88 Cellulose/(CH$_2$O)x/DMSO	1.5-2.8	1.0-1.7	13-18	10-18	0.17-1.0	3-15

duction of rayon fibers from spinning solutions of 6/6/88 cellulose/(CH_2O)x/DMSO composition. The wide range of properties obtained by varying the chemical composition of the regeneration bath affords an appreciation of the nature of the system.

Some fiber physical property data are shown in Table II comparing some commercial rayons with those produced from spinning 6/6/88 cellulose/(CH_2O)x/DMSO solutions into an aqueous 4% sodium sulfide primary regeneration bath.

Conclusions

A wide range of pulps have been found to readily dissolve in the (CH_2O)x/DMSO system including experimental samples. However, groundwood furnish, such as is employed for newsprint will not dissolve in this system. The pulps may be used in either a fluffed, Abbe' cut, shredded or diced sheet form without encountering dissolution problems.

The concentration of pulp which can be used depends upon the degree of polymerization. At 1000 D.P. up to 2% can be spun while at 400-500 D.P. up to 6% solutions were feasible, and at 300 D.P. up to about 10% cellulose solutions could be processed.

The reaction and dissolution process was found to be very specific to DMSO. Analogous solvents were not effective in the presence of paraformaldehyde. Similarly, other aldehydes did not afford cellulose solutions. The cellulose/(CH_2O)x/DMSO solutions were clear and essentially free from gels or fibers therefore requiring only a single stage polishing filtration during spinning.

One unique feature of these fibers is the fact that they exhibit unusually low $S_{6.5}$ values. $S_{6.5}$ values in a range of 3-15% can readily and consistently be obtained. This test is a measure of a fiber or fabric's resistance to alkaline treatment and can be related to launderability performance.

Coagulation and regeneration of the cellulose/(CH_2O)x/DMSO solutions is fairly rapid in the presence of proton donor systems. Fibers with good wet moduli can be spun without the need for spin bath additives.

Conditioned fiber elongation generally was between 4% and 10% which is normally too low for processing. This level could be improved somewhat by proper control of jet stretch to godet stretch ratios. Fiber cross-sectional shapes could not be controlled by the use of various primary bath regenerants and all were circular in appearance.

Comparison of the overall processing steps involved in the DMSO/(CH_2O)x solvent system reveals some distinctive advantages over the complicated viscose process. For example, viscose processing steps such as steeping, pressing, aging, xanthation, ripening and filtration may be eliminated or simplified in a solvent spinning system.

However, it should be pointed out that the commercialization of any solvent process for rayon production will depend to a large extent on the development of a suitable recovery system, ie one that is totally recyclable, economic and non-polluting.

Literature Cited

1. Fowler, W. F. and Kenyon, W. O., J. Amer. Chem. Soc., 69 1636 (1947).
2. Williams, H. D., U.S. Patent No. 3,236,669 (1966).
3. Hergert, H. L. and Zopolis, P. N., French Patent No. 1,469, 890 (1967).
4. Clermont, L. P., (a) Canadian Patent No. 899,559 (1972); (b) Monthly Research Notes, Dept. of Fishery and Forestry, Canada 26, No. 6, 58 (1790); (c) J. Poly. Sci., 10, 1669 (1972), (d) J. Appl. Poly. Sci., 18, 133 (1974).
5. Schweiger, R. G., (a) Chemistry and Industry 296 (1969); (b) U.S. Patent No. 3,702,243 (1972); (c) German Patent No. 2,120,964 (1971).
6. Chu, N. J., Pulp and Paper Research Institute of Canada, Report No. 42 (1970).
7. Pasteka, M. and Mislovicova, D., Cellulose Chemistry and Technology 8, 107 (1974).
8. Hata, K., Yokota, K., J. Sci. Fiber Sci. Techn. Japan 24, 420 (1968).
9. Philipp, B., Schleicher, H., Laskowski, I., Faserforsch Textiltech. 23, 60 (1972).
10. Koura, A., Scheicher, H., Philipp, B., Faserforsch Textiltech. 23, 128 (1972).
11. Johnson, D. C., Nicholson, M. D., and Haugh, F. C., General Paper No. 67 Eighth Cellulose Conference May 19-23, 1975.

Cellulose Ethers and Esters

Cytrel® Tobacco Supplement—A New Dimension in Cigarette Design

R. W. GODWIN and C. H. KEITH

Celanese Fibers Co., Charlotte, N.C. 28232

Over the past 25 years there has been a dramatic decrease in the delivery of "tar," nicotine, and other smoke components from cigarettes. This continuing trend has been particularly marked in the past few years with the introduction of a large number of new brands delivering between 1 and 15 milligrams of tar. These represent a large change from the 25 to 30 milligram brands that were common in the 1940's and 50's. The tobacco industry has brought about this change in response to consumer demand for "lower delivery" cigarettes.

There are a variety of existing techniques which are used to decrease the smoke delivery of cigarettes. Among these are filtration, porous and/or fast-burning papers, filter ventilation, modification of the tobacco filler, and changes in cigarette dimensions.

Filtration is most commonly used and generally achieves up to a 50-60% reduction in tar and nicotine delivery, but it has little effect on gas phase constituents (1). Selective filter media such as charcoal are used in some brands to achieve sizable reductions in some smoke components, while leaving others relatively unchanged. Porous and/or perforated papers are widely used to achieve 10-30% reductions in both gas and particulate phase components. More recently, filter ventilation by means of perforated tipping papers has been employed to achieve another 50 or more percent reduction in all smoke components (2). Finally, changes in the paper burning rate, cigarette dimensions, and the incorporation of expanded tobacco leaf and reconstituted tobacco sheet has been used to create faster burning and lower puff count cigarettes, which directly translates into lower smoke chemistry.

Cytrel® is a registered trademark of the Celanese Corporation.

In general, these techniques are used in combination with each other to achieve a lower delivery cigarette with adequate consumer appeal. However, each of these existing methods has limitations. Some which have been mentioned include the action of filters on the particulate phase of tobacco smoke only, and the across-the-board action of ventilation. Because of these, it is desirable to have additional techniques for controlling the delivery of various smoke components.

One such technique, which has been under development in our laboratories for a number of years, is the inclusion of tobacco supplement in a cigarette. This material, which has the trademarked name Cytrel, is a manufactured sheet material of controllable composition which looks, processes, and burns like tobacco. It has a further desirable property of having a bland taste upon combustion, so that blends of Cytrel with tobacco taste essentially like a mild tobacco. It is the purpose of this paper to to describe the physical, chemical, and some of the pertinent biological properties of this material. The effect of Cytrel inclusion in tobacco blends is also described.

Cytrel is essentially a filled film material, formed by casting and drying a sheet from a dough-like mixture of organic binders and additives with mineral fillers (3). A typical composition is given in Table I. The finished sheet is cut into one-inch squares which are subsequently blended with tobacco and manufactured into cigarettes by conventional techniques. For the purposes of this discussion, cigarettes described are typical of the construction commonly utilized in Great Britain. These cigarettes were 72 millimeters long with a circumference of 25 millimeters and were equipped with a 16 millimeter, dual acetate-paper filter with a tar removal efficiency of 49% (4). Cigarette weights were held constant at 0.98 gram, and pressure drops ranged from 90 millimeters of water at an air flow of 17.5 milliliters per second for the all-Cytrel cigarettes to 116 millimeters for the all-tobacco cigarettes. Blended cigarettes containing 10, 20, and 50% Cytrel were also prepared, and puff counts linearly ranging from 10.2 to 5.8 under standard smoking conditions were obtained as the percentage of Cytrel inclusion increased. This change in puff count only partially accounts for the decreases in deliveries shown later. If it were the only effect, the expected deliveries from a Cytrel cigarette would be 57% of those of a tobacco cigarette. Since deliveries in the range of 0 to 38% are found, it is evident that other factors are involved. From looking at the composition of the material, it is apparent that 70% of Cytrel consists of incombustible material, and a maximum of 30%

Table I: Composition of Cytrel

Material	Amount
Binder (sodium carboxymethyl cellulose)	19%
Inorganic fillers/combustion modifiers	71%
Humectant/plasticizer (glycerol)	5%
Colorants	3%
Flavor additives	2%

can burn during smoking. Taking 30% of 57% gives 17%, which should be the delivery of any given component if no change in combustion occurs.

Before examining the relative deliveries of various components from Cytrel and tobacco, it is necessary to refer to the methods employed and the general pattern of deliveries. The methods utilized and the deliveries of 250 compounds and 75 elements from these cigarettes have been reported in a series of papers by Vickroy, Mauldin, and Allen in the Beiträge zur Tabakforschung (4, 5, 6). Briefly recapitulating, the volatile smoke components, which are those which pass through the glass fiber smoke collection trap, were measured by either gas chromatographic techniques or specific tests for individual components (e. g., hydrogen cyanide, nitric oxide). The semivolatile components are those which can be distilled from the collection pad at a temperature of 130°C. These were analyzed by a combination of gas chromatography and mass spectrometry. The condensed phase components (i. e., those remaining on the collection pad) were measured by a variety of gas chromatographic and specific techniques. The elements were measured by spark source mass spectography except for mercury, which was measured by atomic flourescence spectroscopy. Figures 1, 2, and 3 illustrate the general delivery pattern observed for almost every component, a substantial linear decrease in delivery as the amount of Cytrel in the blend is increased. Figure 1 shows the decline in "tar" which is defined as total particulate matter collected on the filter pad minus water and nicotine. It is evident that there is more than a fourfold decrease in this mixture of materials from 15.8 to 3.6 milligrams/cigarette. Figure 2 is a similar graph for another important smoke component, nicotine. In this case the line goes to zero for 100% Cytrel, as nicotine is absent from the Cytrel formulation. This pattern is characteristic of a number of nitrogen-containing smoke constituents because of the rather low nitrogen content of Cytrel.

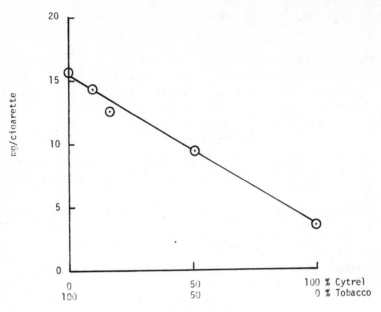

Figure 1. Tar delivery

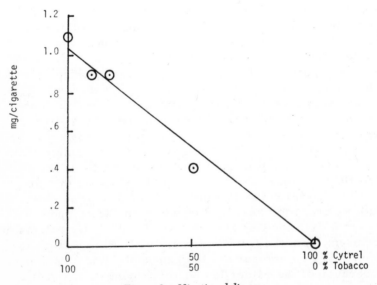

Figure 2. Nicotine delivery

Figure 3 shows a similar delivery pattern for an important gas phase constituent, carbon monoxide.

In Figure 4 we have an example of one of the rare instances where a different pattern is observed. In this case, glycerol increases linearly as the amount of Cytrel in the blend is increased, going from zero in the uncased tobacco cigarette to 2.1 milligrams in the 100% Cytrel cigarette. In cased cigarettes (such as those used in the United States) where glycerol is commonly added as a humectant, this change is not nearly so marked. It is known that glycerol is distilled into the smoke stream as a Cytrel cigarette combusts, and comprises about 58% of the tar collected. Only two other compounds of the 250 examined show an increase in delivery as the Cytrel content in the blend is increased. These are: (1) hydrogen sulfide, which modestly increases from 46 to 55 micrograms/cigarette, and (2) ammonia, which remains constant at 20-22 micrograms/cigarette up to a 50% blend inclusion of Cytrel, but then increases to 50 micrograms/cigarette for an all-Cytrel cigarette. These materials are known to be derived from residual sulfur in the colorant used and decomposition of a nitrogeneous flavor additive, respectively. Among the elements, all but sodium markedly decrease as the Cytrel content of the blend is increased. Sodium, which can be derived from the sodium carboxymethyl cellulose binder, shows as erratic increase with increasing Cytrel content; but this is not nearly as marked as the decrease in potassium, which is a major element in tobacco.

As indicated in the previous discussion, a comparison of smoke component deliveries can best be done considering the compositions of Cytrel and tobacco. It therefore is of interest to examine the relative deliveries of a number of smoke components to see if the overall combustion process can be better understood. From the differences in puff counts and the inorganic content of Cytrel, it is expected that the Cytrel cigarette should deliver 17% of the amount of any component in tobacco smoke if all other factors were equal. In Table II the relative deliveries of gases and vapor phase components are listed as calculated from the Vickroy data (4). For most of these components, the deliveries are near or below the expected 17% delivery (with the exceptions previously noted for ammonia and hydrogen sulfide). The low levels of alkanes, alkenes, benzene, and toluene in Cytrel smoke reflect the absence of saturated hydrocarbons such as tobacco waxes in Cytrel. The relatively higher levels of alkynes, principally acetylene, suggest that free radical reactions are occurring in the pyrolysis process. The very low levels of alcohols, furans,

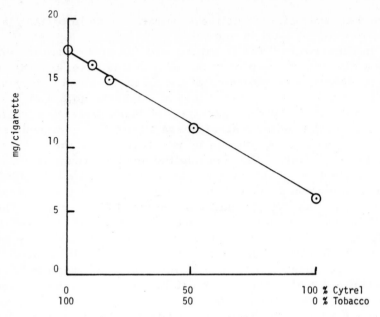

Figure 3. *Carbon monoxide delivery*

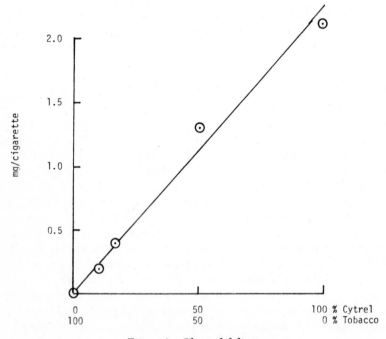

Figure 4. *Glycerol delivery*

Table II: Relative Deliveries of Vapor Phase Components

Material	Number of Compounds	Delivery from Tobacco (μg/cig)	Delivery from Cytrel (μg/cig)	Relative (%)
Hydrocarbons				
C_1-C_6 alkanes	6	2165	312	14.4
C_1-C_6 alkenes	10	1106	169	15.3
C_1-C_5 alkynes	2	48	18	37.5
Benzene	1	104	13	12.5
Toluene	1	163	12	7.4
Methanol and ethanol	2	560	< 4	0.7
Saturated aldehydes (C_2-C_5)	4	< 1817	< 294	16.2
Unsaturated aldehydes (C_3-C_4)	2	164	48	29.3
Ketones (C_3-C_5)	5	1219	238	19.5
Furans (C_4-C_6)	3	287	29	10.1
Nitroles (C_2-C_4)	4	< 315	< 35	11.1
Carbon monoxide	1	17600	6000	34.1
Carbon dioxide	1	67700	48200	71.2
Hydrogen cyanide	1	280	25	8.9
Nitric oxide	1	93	31	33.3
Ammonia	1	22	50	227
Amines (C_1-C_4)	8	< 9.5	< 1.5	15.8
Hydrogen sulfide	1	46	55	119
Sulfur dioxide	1	4.2	1.2	28.6

and nitriles, and the normal and elevated relative levels of aldehydes and ketones, suggest that a somewhat higher degree of oxidation is occurring in the more dilute smoke from the Cytrel cigarette. These oxygenated materials are of course expected from the degradation of the cellulosic binder. The higher relative level of carbon dioxide again suggests a more oxidative combustion process, although some contribution to this component could be expected from a simple thermal decomposition of the carbonate filler material. With the already noted exception of ammonia, nitrogen containing constituents in this phase and elsewhere are generally reduced in Cytrel smoke below the expected delivery levels, primarily because of the low nitrogen level in the formulation. Nitric oxide is an exception to this rule in that its delivery is not reduced as much as might be predicted. From examination of a number of Cytrel formulations, there is a suggestion that this compound may be partially coming from an oxidative degradation of a nitrogeneous flavor additive.

Turning now to the semivolatile fraction of smoke, many materials, particularly those containing nitrogen, are undetectable or absent in Cytrel smoke. From Mauldin's (5) semiquantitative data, relative deliveries can be estimated for several materials which are both detected and resolvable in Cytrel and tobacco smoke. These are presented in Table III. In this table many of the relative deliveries are shown to be below the expected 17% level, indicating the general trend of less complexity in the Cytrel smoke. The substituted pyrazines show somewhat higher relative deliveries than many other components. It is thought that these arise from reactions of ammonia with aldehydes produced by the degradation of cellulose. Similarly, the substituted and unsubstituted cyclopentenones appear to be relatively abundant. These can come from pyrolysis of the anhydroglucose units in the cellulosic binder.

Table III: Relative Deliveries of Semivolatile Components

Material	Delivery from Tobacco (arbitrary units)	Delivery from Cytrel (arbitrary units)	Relative Delivery (%)
Dimethylpyrazine	10	1	10.0
Trimethylpyrazine	14	3	21.4
2-Cyclopenten-1-one	216	17	7.9
2-Methyl-2-cyclopenten-1-one	203	17	8.4
2,3-Dimethyl-2-cyclopenten-1-one	126	22	17.5
2-Ethyl-3-methyl-2-cyclopenten-1-one	16	5	31.3
Tetradecane	37	1	2.7
3-Furfural	40	4	10.0
2-Furfural	143	14	9.8
5-Methylfurfural	96	2	2.1
Methylindene	8	1	12.5
Dimethylindene	31	2	6.5
Acetophenone	38	5	13.2
Napthalene	59	2	3.4
Methylnapthalene	36	3	8.3
Nonadecene	46	2	4.3
1-Indanone	23	3	13.1
3-Methyl-1-indanone	48	1	2.1

Turning now to the less volatile materials in the condensed phase, Table IV presents relative deliveries computed from Allen's data (6). As was evident for the volatile and semivolatile materials, the relative deliveries are close to or below the expected level. The only value apparently above the expected value of 17% is that for "tar." Other analyses on different cigarettes have shown either the expected or below-expected deliveries for tar. As indicated previously, glycerol is a major tar component in Cytrel smoke which is directly distilled out of the formulation. If the glycerol contribution is removed from the tar figure, the relative delivery of the remaining materials decreases to 9.5%. The deliveries of phenolic constituents appears to be greatly reduced. This is an indication that pyrolysis of cellulosic materials does not lead to phenolics in this type of combustion. The probable source of these materials in tobacco smoke is a breakdown of the complex phenolics present in tobacco leaf such as scopoletin, chlorogenic acid, and rutin.

Table IV: Relative Deliveries of Particulate Phase Components

Material	Delivery from Tobacco (per cig)	Delivery from Cytrel (per cig)	Relative Delivery (%)
Tar	15.8 mg	3.6 mg	22.8
Nicotine	1.1 mg	0.0 mg	0.0
Glycerol	<0.1 mg	2.1 mg	--
Phenol	52 µg	< 0.5 µg	1.0
o-Cresol	14 µg	< 0.5 µg	3.6
m- and p-Cresol	27 µg	< 0.5 µg	1.9
Phenanthrene + anthracene	370 ng	22 ng	5.9
3-Methylphenanthrene	200 ng	6 ng	3.0
Flouranthene	130 ng	19 ng	14.6
Pyrene	91 ng	9 ng	9.9
Benzo(a)pyrene	20 ng	4 ng	20.0
Dimethylnitrosamine	21 ng	ND	0.0
Calcium	150 ng	11 ng	7.3
Silicon	100 ng	14 ng	14.0
Arsenic	21 ng	0.07 ng	0.3
Cadmium	250 ng	0.53 ng	0.2
Mercury	0.46 ng	0.028 ng	6.1
Nickel	1.4 ng	0.02 ng	1.4

ND = not detected

The relative deliveries of polynuclear hydrocarbons also appear to be at or below the expected levels. The very small amounts of these materials and the considerable difficulty in their analysis make the relative delivery numbers variable and imprecise. Dimethylnitrosamine is essentially absent in Cytrel smoke. This is not surprising since these compounds are thought to be derived from the reaction of secondary amines with nitric oxide, both of which are present to a lesser extent in Cytrel smoke as compared to tobacco smoke. Calcium and silicon, the metallic elements present in the formulation, have relative deliveries somewhat below the expected level. Other trace elements are greatly reduced in relative delivery.

Thus, for a variety of components in each phase of smoke, the delivery from Cytrel cigarettes is at or below the amount that would be expected from the differing puff counts and inorganic content of the material. In a relatively few instances, greater-than-expected yields are obtained, and these can be related to specific components of the formulation. The generally reduced deliveries for most components suggest that the Cytrel is being more completely combusted into simple molecules (such as carbon dioxide) because of the dilution of the burnable material with inorganics, which provide a considerable surface for pyrolysis.

To further illustrate the unique properties of Cytrel, it is of interest to relate its chemical properties to some observed biological effects. As shown in Table V, there appears to be a correlation between the carbon monoxide content of the smoke and the carboxyhemoglobin levels in animals exposed to the smoke (7, 8, 9). A linear regression of carboxyhemoglobin vs. carbon monoxide for these data gives a line with a slope of 1.42 and an intercept of 7.71. The correlation coefficient is 0.84, which is quite good considering the differences in species under test and the differences in exposure systems.

Table V: Carboxyhemoglobin Levels

			Carboxyhemoglobin (%)*		
			Tobacco	50/50 Blend	Cytrel
	Smoking	Duration of	17.6 mg/	11.5 mg/	6 mg/
Species	Conditions	Exposure	cig CO	cig CO	cig CO
Male and female rats	3% smoke for 1 hr/day	8 weeks	27.4	30.9	14.4
Male hamsters	24 puffs/day of 22% smoke	9 weeks	31.3	19.3	11.3
Male monkeys	10 cig/day	8 weeks	36.7	32.4	18.7
Female monkeys	10 cig/day	8 weeks	30.6	24.4	15.0

*Measured immediately after last exposure

With the greatly reduced levels of irritants in Cytrel smoke, it is to be expected that the material would give a lower response in tests of ciliary transport, macrophage inhibition, and cyto- toxicity (10). These in vitro tests essentially measure the inacti- vation of two important lung defense mechanisms by smoke and the general toxicity of water soluble smoke components towards cell cultures. Table VI lists the results obtained by the National Cancer Institute for tobacco, a Cytrel variant, and a 50/50 blend of the two. In Table VI it is evident that more than eight puffs of Cytrel and blend smoke are required to give a 50% inhibition of ciliary transport, while only 2.5 puffs of tobacco achieve this level of inhibition. Considerably larger quantities of Cytrel and blend smoke are required to inhibit the phagocytosis of bacteria by a suspension of alveolar macrophages by 50%. The cytotoxic properties of Cytrel and, to a lesser extent, blend smoke, are also apparent as only 13 to 18 50% toxic doses are found in a puff of these smokes, whereas 22 such doses are contained in a puff of control tobacco smoke. Further, more extensive biological testing has or is being conducted on Cytrel and tobacco smoke and will be eventually published. The general results are:

1. After acute and chronic inhalation, Cytrel smoke gives a reduced irritating response as measured by established histopathological criteria.
2. Minimal and similar effects are found in reproductive tests on Cytrel and tobacco. These appear to be related to carbon monoxide levels.
3. The tumorigenic activity of Cytrel and blend smokes is reduced when compared to all-tobacco smoke on a per cigarette basis, both in dermal and inhalation test sys- tems.

Table VI: In vitro Biological Responses

Material	Ciliary Inhibition (puffs/ED_{50})	Macrophage Inhibition (ml smoke/ID_{50})	Cytotoxicity ($ED_{50}s$/puff)
Tobacco	2.5	4.4	22.0
50/50 Blend	> 8.0	7.2	18.1
Cytrel	> 8.0	7.8	13.1

With this extensive background of improvements in the chemi- cal and biological properties of Cytrel smoke, it is of interest to estimate what Cytrel inclusion could do for existing types of cigarettes. In this country the most popular brands of cigarettes have a tar delivery of around 17 milligrams, a nicotine delivery of about 1.2 milligrams, and a carbon monoxide delivery of 15 milligrams per cigarette. Replacing 30% of the tobacco with Cytrel should reduce these deliveries to 13.1, 0.8, and 13 milligrams per cigarette, respectively. Thus, a medium tar brand can be changed to a lower tar brand by inclusion of 30% Cytrel. Similarly, a ventilated low tar brand might have tar, nicotine, and CO deliveries of 8.0, 0.6, and 9.9 milligrams per cigarette. Inclusion of 30% Cytrel in the blend could further reduce these to 6.5, 0.4, and 8.6 milligrams per cigarette, respectively.

In conclusion, we have shown that Cytrel tobacco supplement offers a means to reduce deliveries of virtually all smoke components of a cigarette. These reductions are for the most part greater than would be expected on the basis of the burning rate of the material and its composition. The simpler smoke composition and the reduced deliveries of Cytrel-containing cigarettes appear to translate into a reduction in biological response in animal test systems. Inclusion of Cytrel in cigarettes can be readily utilized with other existing techniques to provide lower delivery cigarettes.

Abstract

Cytrel is a tobacco supplement consisting of mineral components, combustion modifiers, and organoleptic-active materials bound together by a film-forming cellulose ether. Cytrel is being used in several brands of cigarettes on the European market at a 20-25% level blended with tobacco. The amount of "tar" contributed to the "mainstream" smoke by Cytrel is one-fourth to one-seventh of that from tobacco. There are similar reductions for other non-nitrogeneous smoke constituents relative to tobacco. Neither Cytrel nor Cytrel smoke contains nicotine, and the number and amount of nitrogen-containing compounds in Cytrel smoke are drastically reduced. Similarly, the amount and complexity of the combustion products in Cytrel "sidestream" smoke are significantly reduced. It has been shown that the delivery of smoke components from blends of Cytrel and tobacco is reduced in direct proportion to the level of Cytrel in the blend with tobacco. This relationship has been demonstrated for virtually all the components of smoke studied to date. Cytrel can be readily blended with tobacco in conventional cigarette making equipment. Further, Cytrel/tobacco blend technology is fully compatible with other means currently being used to modify cigarette smoke, such as various cigarette filters and high porosity cigarette papers.

Literature Cited

1. Keith, C. H. , Proc. 3rd World Conf. Smoking and Health, DHEW Publication No. (NIH) 76-1221 (1976), 1, 49.
2. Norman, V. , Beitr. z. Tabakforsch. (1974), 7, 282.
3. Miano, R. R. , and Keith, C. H. , U. S. Patent 3,931,824 (1976).
4. Vickroy, D. G. , Beitr. z. Tabakforsch. (1976), 8, 415.
5. Mauldin, R. K. , Beitr. z. Tabakforsch. (1976), 8, 422.
6. Allen, R. E. , and Vickroy, D. G. , Beitr. z. Tabakforsch. (1976), 8, 430.
7. Buch, S. et al, Inveresk Research International Report 307, Project 403318 (1975).
8. Bernfeld, P. , and Homburger, F. , Bio-Research Consultants, Inc. Report, Project C-211 (1975).
9. Black, A. et al, Inveresk Research International Report 408, Project 403365 (1975).
10. Gori, G. B. , "Towards Less Hazardous Cigarettes, Report No. 2, The Second Set of Experimental Cigarettes," National Cancer Institute, Smoking and Health Program, Washington, D. C. (1976).

8

Characterization of Insoluble Cellulose Acetate Residues

W. B. RUSSO

Fiber Industries, Inc., Charlotte, N.C. 28237

G. A. SERAD

Celanese Fibers Co., Charlotte, N.C. 28232

Commercial cellulose acetate is normally manufactured using a very high purity wood pulp because its applications have very critical processing and product requirements. Cellulose acetate films must be clear and free of imperfections that could be caused by undissolved particles during solvent film castings. The preparation of cellulose acetate fibers requires the polymer solution to flow unobstructed through spinneret capillaries that have extremely small diameters. Any undissolved particulate matter that would cause disruptions in fiber extrusion cannot be tolerated. Since cellulose derived from wood pulp contains impurities which do not form acetate esters soluble in commercial cellulose acetate solvents (e. g. , acetone), these impurities must be reduced to an acceptable level. Consequently, wood pulps used to manufacture cellulose acetate contain 94 to 99 percent alpha-cellulose, the remainder being hemicelluloses. This α-cellulose level is higher than that required for the preparation of rayon, cellulose nitrate, or cellulose ethers. Of course, the increased purity results in an increased cost. Hence, it has long been a goal to develop means to reduce the wood pulp purity requirements for the commercial preparation of cellulose acetate. This is desirable for the acetate manufacturer, the pulp manufacturer, and the ecology since the higher purity comes at a sacrifice in wood yield and increased pulp mill effluent treatment requirements.

It is well known that insoluble residues obtained from the dissolution of cellulose acetate in acetone are enriched in hemicellulose acetates (1-4). However, our program to investigate the use of lower purity wood pulps to prepare cellulose acetate by either modifying the pulps or the process required quanti-

96

tative characterizations. Hence, this study was undertaken to
isolate, quantify, and characterize both soluble and insoluble
fractions of cellulose acetate in acetone.

Four general classes of pulp, representing various combinations of wood furnish and pulping process, were examined.
These included a softwood kraft, softwood sulfite, hardwood
kraft, and hardwood sulfite. The purity of the various pulps,
along with typical acetylation grade pulps, are given in Table I.

TABLE I: Purity of Acetylation Grade and "Low Purity" Wood
          Pulps

| | \multicolumn{3}{c}{Sugar Analysis (%)} | Alkali Solubility |
	Glucose	Mannose	Xylose	Purity-R_{10}%
"Low Purity" Grades				
Softwood Kraft	85.6	6.1	8.3	84
Softwood Sulfite	91.2	6.9	1.9	85
Hardwood Kraft	83.5	0.6	15.9	89
Hardwood Sulfite	91.5	2.4	6.1	87
Typical Acetylation Grade				
Softwood Sulfite	98.1	1.4	0.5	96
Hardwood Prehydrolyzed Kraft	97.5	0.3	1.2	97

Fractionation Sequence

Acetone, the conventional solvent used to manufacture
cellulose acetate fibers and films, was used to prepare the
starting solutions. The exact solvent composition used was 95/5
(wt/wt) acetone/water. The water includes that contained in both
the solvent and the cellulose acetate flake. A 6 g/100 cc solution
of the cellulose acetate (bone dry basis) was used. The first step
in the fractionation (Figure 1) was ultracentrifugation at 15,000
rpm for four hours. After decanting off the supernate, a fresh
acetone/water solution was added to the residues and ultracentrifugation repeated under similar conditions. The combined
supernates, called the acetone/water solubles (AWS), were
saved for analysis. The residue was weighed and likewise saved
for analysis. A portion of the residue was subsequently prepared as a 6 g/100 cc solution in 91/9 methylene chloride/

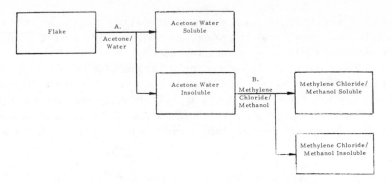

A. 1. 6 gms of "bone dry flake" per 100 cc
    2. Ultracentrifugation (15,000 rpm for 4 hrs)
    3. Decant supernate
    4. Add 150 cc of fresh acetone/water
    5. Repeat steps 2 and 3
    6. Collect liquid, weigh solids

B. 1. Dry residue
    2. Prepare 6% (wt/vol) solution in
        methylene chloride/methanol
    3. Ultracentrifugation (151,000 rpm for 2 hrs)
    4. Decant supernate
    5. Add 150 cc of fresh $MeCl_2$/MeOH
    6. Repeat steps 3 and 4
    7. Collect liquid, weigh solids

Figure 1. Fractionation sequence

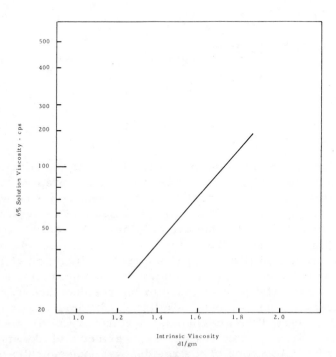

Int rinsic Viscosity
dl/gm

Figure 2. Relationship of cellulose acetate intrinsic
viscosity and 6% solution viscosity

methanol. The solution was ultracentrifuged at 15,000 rpm for
two hours. The supernate was decanted, fresh methylene chlor-
ide/methanol added, and the centrifugation repeated. The combined supernate, designated methylene chloride/methanol soluble
(MCMS), was saved for analysis. The final residue, designated
methylene chloride/methanol insolubles (MCMI), was weighed
and saved.

Analytical Procedures

Acetyl Value (defined as percent combined acetic acid).
A five wt. percent solution of the cellulose acetate flake was
prepared in a 91/9 (wt/wt) methylene chloride/methanol solvent.
A film approximately 0.001 inch thick was cast from the solution
and allowed to air dry. It was placed in a curved film holder,
heated to remove residual water, and scanned in an infrared
spectrophotometer over the 2.5 to 4.5 micron range. The ratio
of the OH absorbence at 2.9 microns and the C-H absorbence at
2.4 microns was calculated. This ratio was correlated to acetyl
value by means of a calibration procedure using a known standard. For a cellulose acetate of 55.07 acetyl value, the sample
standard deviation (ten analyses) was 0.02.

Filtration Value. Moisture content of cellulose acetate
samples was determined by oven drying. A 6 g/100 cc solution
of the sample was prepared using either a 95/5 (wt basis) ace-
tone/water solvent or a 91/9 methylene chloride/methanol solvent. This was shaken for two hours to assure dissolution. The
solution was filtered through 30-ply Kimpak and Canton flannel
at 200 psig nitrogen pressure until the filter was completely
plugged. A filtration value was defined as the grams of dry
cellulose acetate per cm^2 of filter area which can be filtered
before blockage occurs.

Cation Analysis. Calcium, magnesium, and sodium were
determined by atomic absorption using conventional techniques.

Intrinsic Viscosity. Intrinsic viscosity of cellulose acetate
flakes was determined by measuring the viscosity of a 6 g/100 cc
solution of the flake in acetone/water and using the correlation of
Figure 2. Solution viscosity was determined using a Nametre
direct readout visco meter (Model 7-006) at $25°C$.

Carboxyl Content. An ion exchange procedure was used for the determination of the carboxyl content. Flake was treated with HCl to convert the carboxyl groups to their acid form. Carboxyl protons were then liberated by exchange with calcium acetate and subsequently titrated.

Sugar Analysis. Quantification of the carbohydrate content was accomplished by gas chromatography of the trimethyl-silated sugar monomers which were obtained by hydrolysis of the cellulose acetate or wood pulp (5, 6). Special procedures were developed to characterize 70-100 mg samples. An accurate dry sample weight and acetyl value were first obtained and the following hydrolysis procedure used:

1.   The sample was dried at 15-20 mm pressure and $65^{\circ}C$ for about two hours, cooled in a dessicator to room temperature, then weighed.

2.   One ml of 77 percent sulfuric acid was added for solvation and shaken for one to three hours. Extremely discolored samples were discarded.

3.   The solvated sample was diluted with 25 ml of distilled water and 5 ml of a 2.0 mg/ml myo-inositol solution (reference sugar).

4.   The solution was then refluxed for four hours.

Next, a neutralization step consisted of placing a 5 ml aliquot of the hydrolyzed sample in a centrifuge tube, adding 0.7 g of barium carbonate powder, and placing the contents into a vacuum oven at $65^{\circ}C$. After neutralization, coincident with the cessation of bubbles, the mixture was centrifuged (2,000 rpm for 10-15 minutes). The resultant supernate was placed into a 10 ml pear-shaped flask and evaporated to dryness under reduced pressure with a rotary evaporator (bath temperature $25-35^{\circ}C$).

Trimethylsilation was accomplished by adding a 1 ml ampule of TRI-SIL (Pierce Chemical Company) into the flask and permitting it to react for one hour at $50^{\circ}C$. The reaction vial was centrifuged to settle the precipitate. A 1-2 micron sample of supernate was used for gas chromatographic analysis. Gas chromatograph conditions were:

1.   A 40-foot, 1/8 inch O.D. stainless steel tube packed with Dexsil 300 GC (5 percent) on Chromasorb W.

2.   Injection port temperature of $250^{\circ}C$, detector temperature of $280^{\circ}C$.

3.   Program sequence was to inject at $160^{\circ}C$ with a 12 minute hold time, followed by a $1^{\circ}C$/minute program rate to $250^{\circ}C$.

Response factors were determined for a variety of sugars, as shown in Table II.

TABLE II: Response Factors for Gas Chromatography

Sugar		RF
α Glucose		1.23
β Glucose		1.22
γ Glucose		1.22
α, β Xylose		1.24
α, β Mannose		1.24
α, β Ribose		1.24
Arabinose		1.24
Sorbitol/Mannitol		1.00
Methyl Glucose		1.20
Gluconic Acid		1.50
Glucuronic Acid		1.50
Glucuronolactone		1.50
Myo-inositol	define	1.00

Results are reported as the percentage of all sugars that are glucose, mannose, or xylose. This does not give the percentage of the pulp or cellulose acetate that is other than the sugar. For example, if the pulp contained large amounts of resin, lignin, ash, etc., the actual sugar content could be much less than that given by the sugar analysis. With acetate grade wood pulps, the amount of other impurities is extremely low and the sugar analysis can be taken as a good approximation of the sugar content of the pulp.

Molecular Weight Distribution. Tetrahydrofuran has been the most commonly used solvent for determining the molecular weight distributions of cellulose and cellulose acetate by gel permeation chromatography. However, it is a marginal solvent for cellulose acetate having a broad acetyl value range. For cellulose acetate made from wood pulp, a prehump appears (eluting at the solvent front) which represents the highest molecular weight fraction. This prehump has been associated with the amount of hemicellulose in the pulp and has been produced artificially by adding mannan at the start of acetylation of cotton

linters. It was felt that a GPC method that would give reprodu-
cible data using a good cellulose acetate solvent would be of
value. From previous studies of the optical properties of cellulose acetate solutions, it was known that a 91/9 ratio of methylene chloride/methanol was a particularly effective solvent.

Use of a binary solvent in GPC is generally not practiced
because of the extreme sensitivity of the detectors to concentration changes. Initially, switching from tetrahydrofuran to
methylene chloride/methanol produced a baseline drift. The
drift made area calculations precarious, so the following
instrument modifications were made.

1.   Refractometer temperature was reduced to $25^{\circ}C$.

2.   Surge tank and reservoir heaters were removed.

3.   Continuous stirring of the 1.5 litre surge tank was
carried out.

4.   All solvent was distilled and standardized by refractive index and added at one time to the reservoir.

5.   Narrow bore tubing was added to the exit lines of the
detector to maintain a back pressure on the detector.

6.   A stagnant solution of solvent was used in the reference side of the detector during operation to reduce pressure
changes. A fresh solvent was recirculated through the system
daily prior to making measurements. Calibration curves were
run frequently.

The methylene chloride/methanol solvent was obtained by
preparing a 92.7/7.3 (wt/wt) azeotropic distillate and correcting
it to 91/9 by the addition of methanol. The 91/9 ratio was used
since it is a better solvent composition than the azeotrope composition. Refractive index was used to assure the correct composition using a calibration curve of known solvent ratios.

Gel permeation chromatographic columns (3/8 inch by
4 inches) were packed with Waters Associates' STYRAGEL of
1×10^{6}, 3×10^{5}, 1.5×10^{4}, and 1×10^{4} Å porosity. Samples
were run as 0.25 percent solutions in either tetrahydrofuran or
methylene chloride/methanol. Columns were at ambient temperature and the flow rate was 1 ml/minute. Molecular size-
elution time calibration was made with polystyrene standards.
These standards are probably not optimum for this solvent
system.

Although a steady baseline was retained for over a two-
month period, the calibration curve slowly and continuously
shifted. The shift was probably due to changes in STYRAGEL
swelling characteristics, which are related to the pore size and

resolution of the column. Changes in resolution have been minimal, with only a small change at the higher counts (lowest molecular weights).

Typical chromatograms of cellulose acetate in tetrahydrofuran and methylene chloride/methanol are shown in Figure 3. What was significant was the lack of the prehump and the "high" molecular weight tail in the methylene chloride/methanol system sample. This difference was due to the better overall solubility of the sample in methylene chloride/methanol compared to tetrahydrofuran.

The slight change in average molecular size on changing from tetrahydrofuran to methylene chloride/methanol was not felt to be due to the inclusion of the prehump into the normal distribution. It was believed due to the differences in diffusion rates and adsorptive effects of the solute towards the Styragel column and methylene chloride/methanol solvent. Therefore, correlation between GPC parameters and conventional molecular weight measurments should be slightly improved compared to the tetrahydrofuran system. This improvement was due to inclusion of the prehump fraction in the distribution curve. A correlation of the molecular size as determined by GPC in both the tetrahydrofuran and methylene chloride/methanol solvents is shown in Figure 4. In the Discussion section, the GPC results with methylene chloride/methanol will be discussed in terms of relative degree of polymerization, recognizing the relationship normally used with tetrahydrofuran data.

Lignin. The procedure used was a modification of that of Chincholi (7). The sample, initially dried under vacuum, was dissolved in cadoxen to make at least four different concentration solutions (e. g., 1.5-12 mg/liter). The solutions, after being stored overnight in a refrigerator, were transferred to a quartz ultraviolet cell and the adsorption recorded at 190-330 nm. The absorption from a cadoxen reference blank was also measured. Knowing the weight of dry sample, concentration of solutions, the absorptions measured, and the extinction coefficient of lignin, the weight percentage of lignin was calculated from Beer's law. The extinction coefficient of lignin was determined using a wood pulp having a Klason lignin content of 0.14 percent.

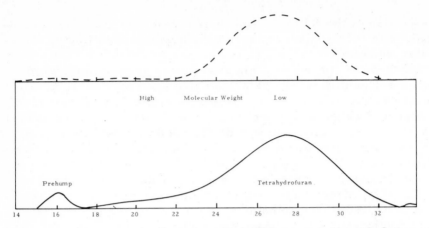

Figure 3. *Gel permeation chromatograph of cellulose acetate in tetrahydrofuran and methylene chloride–methanol*

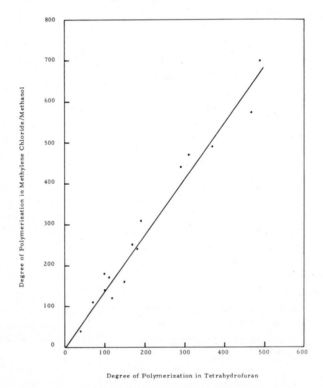

Figure 4. *Comparative degree of polymerization of cellulose acetate in tetrahydrofuran and methylene chloride–methanol*

Cellulose Acetate Preparation

Cellulose acetate was prepared from each low purity wood pulp according to standard practices for the acetic acid-acetic anhydride-sulfuric acid process. Key points included:

1.   The use of a 12-inch Sprout-Waldron disk refiner to fluff the wood pulp sheet.

2.   A 14 percent sulfuric acid catalyst concentration based on bone-dry cellulose.

3.   A Helicone reactor (Atlantic Research Company) for the acetylation and hydrolysis.

4.   Sulfuric acid neutralization with magnesium acetate.

Results and Discussion

Softwood Kraft Pulps. Flake produced from softwood kraft pulps (Table III) had a very low filtration value in the acetone/water solution (< 1 g/cm^2) because of the 30 percent insoluble residue. Although the acetyl value was higher than target (56.3 versus 55.0 target) it was not felt to be a contributor to the poor filterability since acetylation grade pulps produce satisfactory cellulose acetate solutions even at this acetyl value.

The acetone/water soluble fraction, however, had a much improved filterability (23.1 versus <1 g/cm^2 for the original flake). Its average acetyl value was significantly lower (55.2 percent). Sugar analysis showed that about 75 percent of the glucose present in the original flake was present in the acetone soluble fraction. In contrast, only about 40 percent of the mannan and 35 percent of the xylan were present. Hence, there is a preferential exclusion of hemicellulose acetates via acetone solubility. The average degree of polymerization of the soluble portion decreased slightly as compared to the original flake (433 versus 444). There also was a significant decrease in the divalent cation and carboxyl content in the soluble portion. Hence, all of the changes expected with increased solubility in acetone were present.

The acetone/water insoluble residue (30 percent of the original flake) was essentially completely soluble in methylene chloride/methanol. Both the acetyl value (56.6 percent) and degree of polymerization (570) were significantly higher than the acetone/water soluble fraction. The true acetyl value may even be higher than 56.6 percent since the high xylan content (17 percent) tends to reduce the measured acetyl value. This is

TABLE III: Analysis of Cellulose Acetate Prepared from a
Softwood Kraft Wood Pulp

Property	Original Flake	Acetone Soluble Fraction	Acetone Insoluble but Methylene Chloride Soluble Fraction
Weight Fraction, %	100	69.9	30.0
Sugar Analysis, %			
Glucose	80.2	87.8	59.4
Mannose	9.5	5.3	18.6
Xylose	7.1	4.0	17.2
Intrinsic Viscosity	1.69	--	--
Degree of Polymeri-zation	444	433	570
Acetyl Value, %	56.3	55.2	56.6
Carboxyl Content, meq/kg	22.0	11.1	46.0
Cation Content, ppm			
Sodium	60	93	92
Calcium	105	64	157
Magnesium	436	71	965
Lignin, %	--	--	--
Filtration Value, g/cm^2			
Acetone/Water	<1	23.1	--
Methylene Chloride/ Methanol	10	--	--

because the acetyl value as calculated assumes three positions
for acetyl substitution, whereas xylan has only two hydroxyl
positions. Both mannan and xylan enrichment was also confirmed. The material balance shows about 60 percent of the original
mannan and 65 percent of the original xylan remained in this
acetone insoluble residue. Although the cation analysis did not
produce an exact material balance, it did show a marked enrichment in the divalent cations. Compared to the original flake,
the magnesium content doubled and the calcium content increased
50 percent. This is also consistent with the 100 percent increase
in carboxyl content.

The residue insoluble in methylene chloride/methanol
(approximately 0.1 percent) was too small for detailed analysis.

Softwood Sulfite Pulp. The softwood sulfite pulp (Table IV)
also produced a cellulose acetate flake having poor filtration values (<1g/cm^2). Again, there was about a 30 percent insoluble
residue in the acetone/water solution which accounted for about
25 percent of the glucose present in the original flake. However,
in contrast to the softwood kraft, there was also a higher ratio
of mannan to xylan content in the flake than in the initial wood
pulp.

The acetone/water soluble portion of the flake had a higher
glucose content than the original flake. Whereas the acetone/
water soluble portion represented approximately 70 percent of
the original flake, it contained approximately 75 percent of the
original glucose content. The mannan and xylan content represented only 42 percent and 46 percent respectively of that present
in the initial flake. Hence, as with the softwood kraft, there is
an enrichment in glucose and rejection of hemicellulose components in the acetone soluble portion. As expected, the filtration
quality of the acetone/water soluble portion was greater than the
original (< 1 g/cm^2 compared to 20.2 g/cm^2). There was also
a reduced acetyl value, carboxyl content, and average degree of
polymerization in the soluble fraction. There was one anomaly
in the divalent cation content. Although the calcium level dropped
as anticipated, the magnesium level increased. This is believed
in error since it is not consistent with the reduced carboxyl content or the magnesium ion material balance.

The 30 percent fraction of the flake that was insoluble in
acetone was completely soluble in methylene chloride/methanol.
Sugar analysis confirmed an increased hemicellulose level. Both
mannan and xylan levels were approximately double that in the
original flake. All the properties of the acetone insoluble portion

TABLE IV: Analysis of Cellulose Acetate Prepared from a
Softwood Sulfite Wood Pulp

Property	Original Flake	Acetone Soluble Fraction	Acetone Insoluble but Methylene Chloride Soluble Fraction
Weight Fraction, %	100	70. 8	29. 1
Sugar Analysis, %			
Glucose	87. 7	92. 2	72. 3
Mannose	10. 4	6. 0	20. 1
Xylose	1. 0	0. 8	2. 5
Intrinsic Viscosity	1. 64	--	--
Degree of Polymerization	378	334	420
Acetyl Value, %	55. 5	55. 3	56. 3
Carboxyl Content, meq/kg	15. 9	7. 3	14. 8
Cation Content, ppm			
Sodium	69	149	145
Calcium	116	90	197
Magnesium	57	275	354
Lignin, %	0. 34	0. 63	0. 70
Filtration Value, g/cm^2			
Acetone/Water	1	20. 2	--
Methylene Chloride/ Methanol	37	--	--

had changed from the original flake consistent with hindered solubility. There was an increase in acetyl value, carboxyl content, average degree of polymerization, and divalent cation content.

Hardwood Kraft Pulp. This pulp produced a cellulose acetate flake with significantly different solubility characteristics than either of the softwood pulps (Table V). The pulp and the resultant flake are high in xylan content, reflecting the characteristics of the hardwood furnish. There was only a one percent mannan content in the flake, whereas the xylan content was 16.7 percent. The intrinsic viscosity and the average degree of polymerization was higher than flakes from the two softwood pulps. This is merely a consequence of being closer to the target intrinsic viscosity. Approximately 80 percent of the flake was soluble in the acetone/water solution. Even though this was approximately 10 percent more than flake from the softwood pulps, the slight increase in filterability (1.2 g/cm^2 versus 1 g/cm^2) was not of practical significance.

The acetone/water soluble portion of the flake contained an increased glucose content and a reduced mannan and xylan content. Approximately 90 percent of the glucose in the original flake was in the acetone/water soluble portion, whereas only about 60 percent of the mannose and 30 percent of the xylose was present. Hence, there was a significant xylose rejection in the acetone/water soluble fraction. The relatively small, total mannose content actually had little impact. As with the other pulps, there was a reduction in carboxyl content, average degree of polymerization, and divalent cation content in the acetone/water soluble portion. The filterability, as expected, also increased to a respectable 37.5 g/cm^2.

Behavior in the methylene chloride solution also was different from cellulose acetate produced by the softwood pulps. Whereas all the acetone/water insoluble fraction was soluble in methylene chloride using the softwood pulps, only about two-thirds was soluble with the hardwood kraft. The methylene chloride soluble fraction represented only 13.2 percent of the original flake, and the insoluble portion represented 6.3 percent. The methylene chloride soluble portion contained a very low glucose content (27.4 percent) and the remainder of the mannose. The xylose content was approximately 62 percent. There was a further increase in divalent cation content and lignin content. The methylene chloride insoluble fraction had an even further decrease in glucose content with an accordant increased xylose

TABLE V: <u>Analysis of Cellulose Acetate Prepared from a Hardwood Kraft Wood Pulp</u>

Property	Original Flake	Acetone Soluble Fraction	Acetone Insoluble but Methylene Chloride Soluble Fraction	Methylene Chloride Insoluble Fraction
Weight Fraction	100	80.5	13.2	6.3
Sugar Analysis, %				
Glucose	79.2	90.0	27.4	10.0
Mannose	1.0	0.7	3.0	0.1
Xylose	16.7	5.8	61.9	81.8
Intrinsic Viscosity	1.74	--	--	--
Degree of Polymerization	622	500	--	--
Acetyl Value, %	55.6	55.3	--	--
Carboxyl Content, meq/kg	40.2	12.7	144	159
Cation Content, ppm				
Sodium	72	96	62	74
Calcium	523	254	1836	2326
Magnesium	93	59	223	245
Lignin, %	0.53	0.38	1.12	1.60
Filtration, g/cm^2				
Acetone/Water	1.2	37.5	--	--
Methylene Chloride/Methanol	4	--	--	--

content. The carboxyl content of both the soluble and insoluble methylene chloride fraction was extremely high at 144-159 meq/kg.

Hardwood Sulfite Pulp. This pulp was quite distinct from the other three. Whereas the others produced a cellulose acetate having a filtration value no greater than 1 g/cm^2, this pulp gave a flake with 11 g/cm^2 filtration value. The relatively good performance of this pulp was consistent with previous literature ([2], [3]). The sulfite process degrades the xylan but does not linearize it. The hardwoods inherently have a lower mannan content. Hence, a hardwood sulfite pulp exhibits the least effect of hemicellulose content on acetylation product performance. The general commercial availability of pulps of this type, however, are quite limited since they usually have poorer yields for the pulp manufacturer.

Detailed chemical analysis of the insoluble portions was not attempted since the acetone insoluble portion ($<$ 2 percent) was too small for analysis.

Sugar Distribution - Pulp to Fiber. Cellulose acetate prepared from a low purity softwood sulfite was used to prepare fiber using a 91/9 methylene chloride/methanol solvent system. The purpose was to follow the sugar content distribution in pulp, flake, fiber, and filtration residues.

As expected, there was a significant drop in xylan content in going from pulp to flake (Table VI). The sulfite pulping process degrades xylans, allowing them to become more soluble in the acetylation process. The flake was dissolved in methylene chloride/methanol (91/9) and filtered through a compressible fabric medium. The residue filtered out had the same sugar analysis as the original flake. Hence, filtration did not preferentially remove a hemicellulose fraction. Fiber prepared from the solution also had a composition similar to the original flake. Fiber physical properties were comparable to that made with standard acetylation grade wood pulps.

Lignin Contents. The procedure for lignin content was used with cellulose acetate prepared from the softwood sulfite and hardwood kraft pulps. Since the method was based on UV absorption at a specific wavelength, it is possible to have interference from other compounds. Hence, although the results were correlated with Klason lignin, they may not truly be lignin but will be called so. In any case, the two low purity pulps produced flake

TABLE VI: Sugar Analysis at Various Product Stages Using
a Softwood Sulfite Wood Pulp*

Component	Pulp	Cellulose Acetate Flake	Cellulose Acetate Yarn	Filtration Residue
Glucose, %	90.9	88.6	89.9	88.7
Mannose, %	6.1	8.5	7.7	8.2
Xylose, %	2.0	0.8	0.8	0.9

*Note: This is a different softwood sulfite pulp than used in
the fractionation study.

with lignin contents of 0.34 percent (softwood) and 0.52 percent
(hardwood). The lignin results for the softwood were not consistent with the fractionation sequence material balance but they
were for the hardwood pulp. The acetone/water soluble portion
of the latter had a lower lignin content, whereas the insoluble
portion had a significantly higher level.

The lignin procedure was also applied to a series of other
pulps and flakes (Table VII). Acetylation grade pulps and acetate
flakes prepared from them were approximately equal (0.36 to
0.42 percent). Flake prepared from a lower purity pulp was
quite similar. To be sure the method was capable of making
distinctions, a plastics grade pulp and a very impure pulp
(Kappa number of 45) were evaluated. The former was 0.15 percent, whereas the latter was 7.7 percent. Hence, it is felt that
the results for the acetylation and lower purity pulps are consistent.

Conclusions

Cellulose acetate, prepared from cellulose having a lower
purity than acetylation grade wood pulps, is known to produce
acetone solutions of unsatisfactory quality. Properties of both
the soluble and insoluble portions of cellulose acetate in acetone
have been quantitatively characterized.

Softwood pulps, prepared by either the kraft or sulfite process, produce approximately a 30 percent insoluble residue in
acetone solutions. The residue is almost completely soluble in

TABLE VII: Comparative "Lignin" Content

Sample	Weight Percent "Lignin" (Based on Pulp Weight)
Hardwood, Prehydrolyzed Kraft Pulp (Acetylation Grade)	0.38
Softwood, Sulfite Pulp (Acetylation Grade)	0.37
Cellulose Acetate Flake from a Softwood Kraft Pulp	0.36
Cellulose Acetate Flake from a Hardwood, Prehydrolyzed Kraft Pulp (Acetylation Grade)	0.42
Softwood Sulfite Pulp (Plastics Grade)	0.15
Kappa 45 Wood Pulp	7.7

a methylene chloride/methanol solution. Hardwood kraft pulps have a 20 percent residue, of which only two-thirds is soluble in methylene chloride/methanol. The remaining one-third is primarily insoluble linearized xylans. Hardwood sulfite pulps are rather unique in that the cellulose acetate produced has a relatively high acetone solubility (> 98 percent).

Sugar analysis indicates an enrichment in glucose (5 to 15 percent) and a reduction of mannose (30 to 40 percent) and xylose (20 to 65 percent) in the acetone soluble portion of the cellulose acetate.

A gel permeation chromatography procedure to provide a molecular size distribution using a methylene chloride/methanol solvent system was developed. The average degree of polymerization of the acetone soluble portion of the flake was lower (2 to 20 percent) than the original flake.

The acetone soluble portion of the flake prepared from each wood pulp also showed the following identical patterns:

1. Lower average acetyl value (0.2 to 1.1 units lower)
2. Lower calcium ion content (20 to 50 percent lower)
3. Lower carboxyl content (50 percent lower for the softwood and 75 percent lower for the hardwood)

The apparent lignin content of pulp and the cellulose acetate prepared from it are similar. However, the acetone soluble portion of the flake has a lower lignin content.

Both cellulose acetate flake and fiber prepared from a low purity wood pulp retain similar sugar contents. Fiber physical properties, obtained by extruding a methylene chloride/methanol solution, are similar to that obtained with an acetylation grade wood pulp.

Acknowledgement

The authors would like to acknowledge Dr. Wendell Burton, who developed several of the necessary analytical techniques which made the quantitative determinations possible.

Literature Cited

1. Steinman, H. W., and White, B. B., "Mannan in Purified Wood Pulp and Its Relation to Cellulose Acetate Properties," Tappi 37 (6): 225-232, 1954.

2. Wilson, J. D., and Tabke, R. S., "Influences of Hemicellulose on Acetate Processing in High Catalyst Systems," Tappi Dissolving Pulp Conference Atlanta, pp. 55-68, October, 1973.

3. Gardener, P.E., and Chang, M. M., "The Acetylation of Native and Modified Hemicelluloses," Tappi Dissolving Pulp Conference, Atlanta, Georgia, pp. 83-97, October, 1973.

4. Neal, J. L., "Factors Affecting the Solution Properties of Cellulose Acetate," Journal of Applied Polymer Science, Vol. 9, p. 947 (1965).

5. Pettersson, G., "Gas Chromatography--Mass Spectrometry of Sugars and Related Hydroxy Acids as Trimethylsilyl Derivatives," Svensk Papperstidning nr 1, p. 27, 1975.

6. Matsuzaki, K., and Ward, Jr., K., "The Monoglucose Sugar Units in Cellulose Acetate," Tappi 41 (8), pp. 396-402, 1958.

7. Chinchole, P.R., "Residual Lignin in Dissolving Grade Pulp," Tappi Dissolving Pulp Conference, Atlanta, pp. 24-26, October, 1973.

Acetylation of Homogeneously Sulfated Cellulose

W. B. RUSSO and G. A. SERAD

Celanese Fibers Co., Charlotte, N.C. 28232

The most common method for the preparation of cellulose acetate involves the heterogeneous reaction of cellulose with an acetic anhydride-acetic acid solution in the presence of a sulfuric acid catalyst. The heterogeniety of the reaction is reflected in that the interior regions of the cellulose become available to the reagents only as accessible regions in the cellulose react and become soluble. A consequence is that to prepare a useful cellulose acetate the cellulose must be completely esterified to a soluble triacetate and then subsequently hydrolyzed to the desired acetyl value.

A desirable alternative to prepare cellulose acetate would be to dissolve the cellulose in a suitable solvent, react it homogeneously to the desired acetyl value, and quench the reaction to give the product. Unfortunately the traditional solvents for cellulose do not lend themselves to direct acetylation.

Recently a dimethyl formamide/dinitrogen tetroxide solvent system was reported for cellulose (1). It was also reported that cellulose could be sulfated to give a water soluble, homogeneous cellulose sulfate in this solvent. With this former work as a basis, a program was initiated to prepare secondary cellulose acetate using a cellulose solvent system.

Results and Discussion

Direct Acetylation of Dissolved Cellulose. Based on the published reports (1), a variety of wood pulps were dissolved by bubbling N_2O_4 gas into a dimethyl formamide - pulp slurry. A clear viscous solution and color from light yellow to blue-green indicated complete solutioning.

Attempts were made to acetylate the dissolved cellulose directly using a variety of reagents (see Table I). The reagents were used with and without added catalysts, at room temperature and at elevated temperatures. In all cases the regenerated material showed no evidence of acetylation. As a final attempt,

cellulose solutions were treated with ketene. Although color
changes suggested reaction, no evidence of acetylation was
observed after regeneration. We believe that none of these
reagents was capable of replacing the nitrite groups attached
to the cellulose to form an acetate. Instead, it was decided
to concentrate efforts on the regenerated cellulose to circum-
vent this problem.

Table I: Attempted Acetylation of DMF-N_2O_4

Reagent	Result
Acetic Anhydride	No Reaction
Acetic Anhydride/Sulfuric Acid	No Reaction
Acetic Anhydride/Pyridine	No Reaction
Sodium Acetate	No Reaction
Potassium Acetate	No Reaction
Acetyl Chloride	No Reaction
Ketene	No Reaction

Acetylation of Regenerated Cellulose. The accessibility
of the cellulose to reagents is a major consideration in the
preparation of cellulose acetate. A study was made to determine
the combined effects of dissolution of cellulose in a DMF-N_2O_4
solution and regeneration as a pretreatment for acetylation.
An acetate grade wood pulp was dissolved in a DMF-N_2O_4 solution
and then regenerated with acetic acid. It was further solvent
exchanged with acetic acid to remove residual DMF-N_2O_4 by
products. Acetylation of the regenerated pulp at a low consistency (2) (2% solids) was much faster than for the same pulp
pretreated under standard conditions. However, the turbidity
of the final solution was much higher (see Figure 1). This
indicated that although the rate of acetylation substantially
improved, the triacetate product quality was poorer. Thus,
merely increasing the accessibility of the cellulose to reagents
is not sufficient to produce a good acetylation product. Since
this was felt not to be a viable route to preparing useful
cellulose esters, it was decided to explore more uniform sulfation of the cellulose.

Acetylation of Regenerated Cellulose with Homogeneous
Sulfation. The mechanism for the dissolution of cellulose in
DMF-N_2O_4 is reported to be due to the formation of a nitrite
ester. It has also been reported that sulfur trioxide is capable
of displacing the nitrite groups to form cellulose nitrite
sulfate, with cellulose sulfate obtained after regeneration (1).
The high solubility of this regenerated material in water was

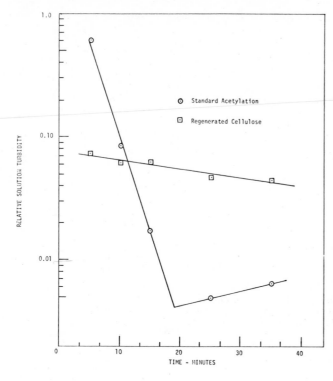

Figure 1. Comparative acetylations of an acetylation grade wood pulp before and after regeneration from DMF–N$_2$O$_4$

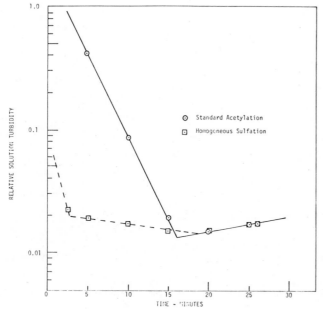

Figure 2. Comparative acetlyations of an acetylation grade wood pulp processed with standard and homogeneous sulfation procedures

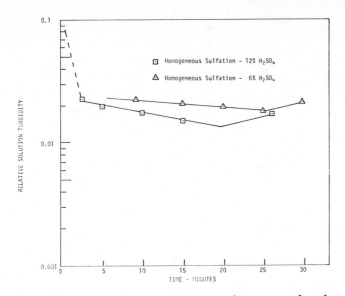

Figure 3. Comparative acetylations of a homogeneously sulfated, acetylation grade wood pulp having different equivalent sulfuric acid concentrations

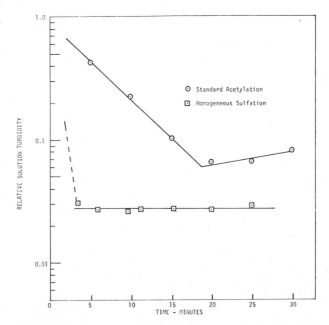

Figure 4. Comparative acetylations of a softwood sulfite pulp processed with standard and homogeneous sulfation procedures

cited as evidence of the homogeneous nature of the sulfate distribution. The acetylation behavior of cellulose sulfate prepared in this manner became the main focus of this study.

To provide samples for acetylation, an acetate grade wood pulp was dissolved in DMF-N_2O_4 and treated with sufficient sulfur trioxide (3) to provide an equivalence of 6% or 12% sulfuric acid. These samples were regenerated with acetic acid to form a gelatinous solid, then solvent exchanged with additional acetic acid. The gelatinous solid obtained was acetylated at a 2% consistency using a standard acetylation test (except for the omission of sulfuric acid.) The reaction was very rapid and gave a final solution turbidity normally associated with good filterability. Figure 2 shows a comparison of the same pulp acetylated by the homogeneous sulfation procedure (12% equivalent sulfuric acid) and acetylated with standard procedures. The rate of esterification of the pulp homogeneously sulfated (containing a 6% sulfuric acid equivalency) was also very rapid (see Figure 3). It was observed that the viscosity of the final cellulose triacetate was unusually high, when either a 6 or 12% equivalent sulfuric acid level was used. Thus it appeared that acetylation grade wood pulps could be acetylated to an equivalent product quality. However, the acetylation occured more rapid than by the traditional process and it occurred with less degradation of the pulp intrinsic viscosity.

Acetylation via the homogeneous sulfation process was examined with several pulps having purities below that of a dissolving grade. Properties of the pulps and cellulose acetate flake prepared from them by conventional procedures have been described previously (4). A softwood sulfite pulp (α-cellulose, ∿85%) showed improved acetylation performance (see Figure 4). The acetylation time was substantially reduced (from eighteen minutes to four minutes). The solution turbidity was lower and increased only slightly with time, in contrast to the same pulp acetylated by standard procedures. In fact, it performed only slightly poorer than the acetylation pulp prepared by the homogeneous acetylation conditions. A softwood kraft pulp (α-cellulose, ∿80%) also showed a markedly improved acetylation performance (see Figure 5). There was a large reduction in solution turbidity as compared to conventional acetylation. The reactivity was also significantly improved. The solution turbidity was better than that obtained using the softwood sulfite pulp.

A hardwood kraft pulp showed an unusual result (see Figure 6). The reaction of this type of pulp with conventional acetylation conditions is slow and the reaction mixture remains quite turbid. The homogeneous sulfation procedure increased the rate of acetylation as with the other pulps and led to a much lower initial turbidity. However, as the reaction continued the turbidity suddenly increased rapidly. The high xylan fraction of the hardwood pulp, linearized by the pulping process crystallized from solution leading to an observable turbidity.

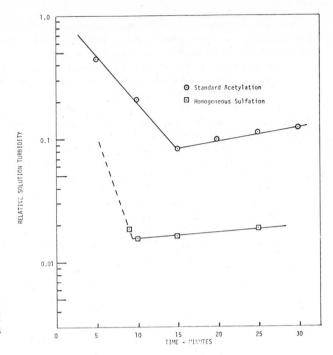

Figure 5. Comparative acetylations of a softwood Kraft pulp processed with standard and homogeneous sulfation procedures

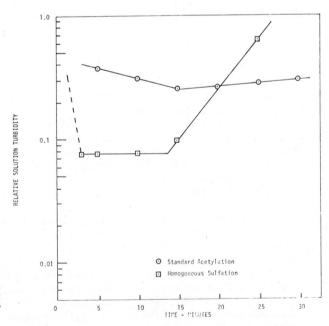

Figure 6. Comparative acetylations of a hardwood Kraft pulp processed with standard and homogeneous sulfation procedures

This is similar to the behavior of acetylated hardwood xylans
described by Gardner and Chang _(5)_ and consistent with the
significant qualtity of xylans found in insoluble acetate
residues with acetone and methylene chloride _(4)_.

The low consistency acetylation test is excellent for
predicting acetylation performance at higher consistency for
acetylation grade wood pulps. However, because of the high
turbidity levels with lower purity pulps, its value for these
pulps is sometimes questionable. In addition, it is difficult
to isolate either a secondary or triacetate product from the
reaction for further characterization. For this reason further
evaluations were conducted with a small-scale, high consistency
(6% solids) acetylation procedure to produce a secondary acetate
flake product. A hydrolysis step was included to reach a target
55.0% combined acetic acid.

An acetylation grade wood pulp was used in the initial
attempts to prepare a secondary acetate flake using the homogeneous sulfation process. Flake was prepared using standard
procedures for hydrolyzing the triacetate to a secondary acetate
but the results were not successful. The acetyl values (combined
acetic acid) were consistently lower than expected. Longer
acetylation times did not improve the situation. A reduction
in the hydrolysis time was successful in increasing acetyl
value. However, the hydrolysis time required to reach target
acetyl value was so short that little more than desulfation
occurred. Hence, based on these data, it was concluded that
a secondary acetate was prepared directly. The solution
properties of this flake in acetone were quite good (see Table II)
compared to that obtained by standard acetylation procedures
in the same equipment.

An experiment was conducted using DMF as a pretreating
agent to activate the pulp prior to conventional acetylation.
This was done to assure that it was the homogeneous sulfation
process that was actually responsible for the improved acetylation and not the DMF pretreatment. A sample of the softwood
sulfite pulp was pretreated by slurrying it in DMF, solvent
exchanging with acetic acid, and acetylating by the normal
process using sulfuric acid as catalyst. A second sample was
dissolved in DMF-N_2O_4, regenerated as cellulose, and acetylated
using sulfuric acid as the catalyst. In both cases, the products
obtained were insoluble in acetone and could not be evaluated.
Thus, as in the low consistency acetylations, making the cellulose accessible was not the major factor in improving the
acetylation.

Further evaluations with the homogeneous sulfation technique
were made with the softwood sulfite and the softwood kraft
pulps which performed well in low consistency tests. Acetylation
of either pulp under standard conditions has always provided a
flake which gave cloudy dope solutions and which blocked
filtration media very rapidly _(4)_. Acetylation of the regenerated cellulose sulfate prepared from these pulps provided

Table II

Cellulose Acetate Flake Properties

Small Scale, High Consistency Acetylation

Grade	Pulp			Acetate Flake Properties		
	Wood	Process	Acetylation Process	Combined Acetic Acid %	6 Weight % Solution Viscosity – cps	Filterability in 95/5 Acetone/Water gm/cm²
Acetylation	Softwood	Sulfite	Standard	55.5	101	8
Acetylation	Softwood	Sulfite	Homogeneous Sulfation	54.1	143	103

Table III

Cellulose Acetate Flake Properties

Small Scale, High Consistency Acetylation

Pulp				Acetate Flake Properties		
Grade	Wood	Process	Acetylation Process	Combined Acetic Acid %	6 Weight % Solution Viscosity – cps	Filterability in 95/5 Acetone/Water gm/cm²
Low Purity	Softwood	Sulfite	Homogeneous Sulfation	54.2	103	75
Low Purity	Softwood	Kraft	Homogeneous Sulfation	54.5	62	56

cellulose acetate flake which was almost comparable to that obtained from the acetylation pulp (see Table III).

Conclusions

It has been shown that it is possible to prepare an acetone soluble secondary cellulose acetate directly without the need to first prepare cellulose triacetate. To accomplish this, it is necessary to distribute the catalyst uniformly on the cellulose. This is readily accomplished by dissolving the cellulose in $DMF-N_2O_4$ and adding sufficient sulfur trioxide. This technique is applicable to a wide range of α-cellulose content wood pulps. For an acetylation grade pulp, the rate of reaction is increased probably as a result of making the cellulose highly accessible to the reagents. However, the filtration properties of the cellulose acetate were similar to that obtained by conventional acetylation techniques.

For low purity pulps (α-cellulose, 80-90%), there is a dual advantage. First, acetylation rate is more rapid due to the accessibility of the cellulose. Secondly, improved solution properties result from the uniform distribution of the sulfate catalyst. It is believed that in the standard acetylation procedure a disproportionate amount of the catalyst is absorbed by the hemicelluloses, leaving the cellulose with insufficient catalyst for a good reaction. This is overcome by distributing the catalyst uniformly on the dissolved cellulose.

Literature Cited

1. Schweiger, R. G., Tappi Dissolving Pulp Conference (1973) Preprints, 167-174.
2. Rosenthal, A. J., International Symposium on Dissolving Pulps Helson K (1966), Symposium Lectures 535-549.
3. Schweiger, R. G., Carbohydrate Research (1972) 21 225.
4. Russo, W. B., and Serad, G. A., Characterization of Insoluble Cellulose Acetate Residues, presented at the 173rd ACS National Meeting, Cellulose, Paper, and Textile Division, New Orleans, La., March 1977.
5. Gardner, P. E., and Chang, M. Y., Tappi Dissolving Pulp Conference (1973) Preprints, 83-96.

Acetylation Behavior of Hydroxyethylated Wood Pulp

JESSE L. RILEY

Celanese Fibers Co., Research and Development Dept.,Charlotte, N.C. 28232

Not all cellulose sources are suitable for acetylation. When conventional acetylation procedures are followed, less highly purified wood pulps containing more than 4% hemicellulose do not produce satisfactory products. The higher hemicellulose content of these pulps results in acetone solutions with higher "false viscosities," dopes with poorer filtration characteristics, and end-products that are yellower.

On the other hand, increases in the prices of acetylation-grade pulps over the past five or six years provide a strong incentive for using lower cost, less highly purified wood pulps. A possible solution to this ongoing dilemma of quality versus price involves structurally modifying lower grade pulps to produce treated pulps that acetylate satisfactorily.

Since a number of alterations in the properties of cellulose and cellulose esters are reported to result from hydroxyethylation, the feasibility of modifying and improving low-purity wood pulps by hydroxyethylation with ethylene oxide was investigated.

Experimental

The hydroxyethylated pulps were prepared by adding 2.5, 5.0, and 7.5% ethylene oxide (on an oven-dried basis) to a nitration-grade (~8-9% hemicellulose) pulp solution containing 50% pulp in 2% sodium hydroxide. The reaction time was one hour at ambient temperature. The estimated conversion efficiency of ethylene oxide to a hydroxyethyl substituent on the pulp was 50%; the rest of the ethylene oxide reacted with water to form ethylene glycol.

A laboratory-scale acetylation procedure developed by Celanese Fibers Company to reliably simulate production-scale acetylation was used to characterize the hydroxyethylated pulps along with control (untreated) nitration-grade pulps. In this test procedure, glacial acetic acid (35% by weight) is added to a 4.00-gram sample of pulp; the pulp is usually conditioned to 6% moisture but it can be bone-dry. The pulp solution is then placed in a tightly sealed 1-liter polyolefin container equipped with

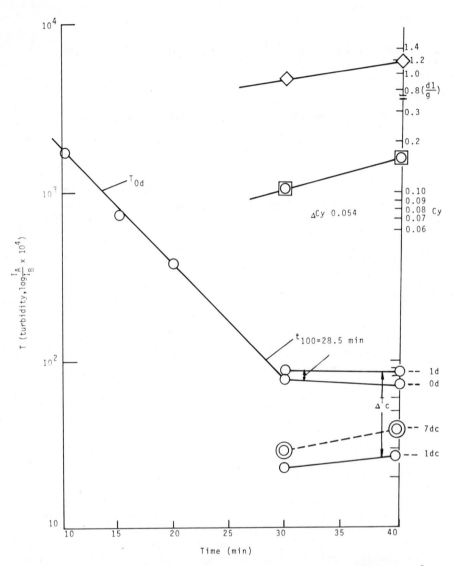

Figure 1. *Turbidity, yellowness coefficient, and intrinsic viscosity measurements characterizing the bench acetylation of a nitration-grade pulp that was moisture conditioned to 6.4%, hammered during pretreatment, and modified by the addition of 2.5% ethylene oxide. Turbidity and absorbance measurements are for a 1-cm cell length at a solids concentration of 2 gm pulp/dl. Turbidity and absorbance are expressed throughout in arbitrary units 10^4 times the actual value. The yellowness coefficient, Cy, is determined on the supernate at 2 gm pulp/dl and a cell length of 5 cm.*

$$\text{Cy} = 1 - (A\ 660\ m\mu / A\ 440\ m\mu) - {}^1\!/_3 \log A\ 660\ m\mu$$

eight Teflon-coated hammers and moderately agitated for one
hour at ambient temperature and pressure. Following this pretreatment, a bench acetylation is conducted at $50^\circ C$ with a total
system volume of 200 ml. An excess of acetic anhydride similar to
that used in commercial operations is added, and the level of sulfuric acid catalyst is 14%. Samples are withdrawn from the
reacting acetylation system at 5- to 10-minute intervals, enough
water added to quench the reaction and increase the water content
to 10%, and characterized optically using a Lumetron 402-E colori-
meter; turbidity and absorbance* are measured on just-agitated
samples so as not to lose the contribution of undissolved fibers.
A minimum or a low plateau (see Figures 1 and 2) is observed in
both turbidity and absorbance as the reaction proceeds. This
minimum (the inception of the plateau) corresponds to reaching
"peak" temperature in an adiabatic commercial acetylation. In the
present study, the minimum usually occurred approximately 30
minutes into the reaction.

Acetylation product evaluations are based on two samples, one
taken near peak and one taken 10 minutes later. Turbidity (T) and
absorbance (A) are measured immediately after sample collection
(T_{0d}) and again one day later (T_{1d}). The samples are then centrifuged, and the supernatant fluids are carefully separated. Immediately after centrifugation, the turbidity (T_{1dc}), absorbance
(A_{1dc}), yellowness coefficient (Cy), and intrinsic viscosity
$([y])$, of the supernatant fluids are determined. Finally, the
turbidity and absorbance of the supernatant fluids are measured
again six days later $(T_{7d}$ and $A_{7dc})$. Highly purified pulps yield
acetylation products which show complete stability over time, but
haze increases during periods of standing at ambient temperature
in the acetylation products of less pure pulps.

The relationship between the turbidity of samples collected
during bench acetylation and reaction time is log linear because
acetylation is a first-order reaction when acetic anhydride is
present in excess (i.e. is not rate-limiting). The rate-deter-
mining step appears to be the reaction at the crystalline cellu-
lose/liquid medium interface which forms a soluble acetic/sulfuric
triester. Although one might assume that the accessible surface
cellulose would be proportional to an exponential function of mass,
it is, in fact, linearly related to mass as shown by the pro-
portionality of turbidity to the amount of unacetylated fibers.

---

*The turbidity, T, is here defined as $\log I_A/I_B$ where I_A is
intensity of transmittance with the optical cell adjacent to the
detector, I_B is the corresponding intensity with the cell adjacent
to the source. The absorbance, A, is defined as $\log I_S/I_A$ where
I_S is the intensity of transmittance for the solvent in the cell
position adjacent the detector and I_A is as foregoing.

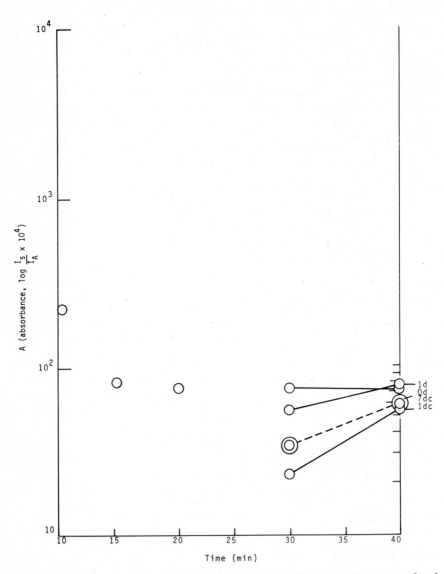

Figure 2.   Absorbance measurements characterizing the bench acetylation considered in Figure 1

Results

The results of a bench acetylation of a nitration-grade pulp treated with 2.5% ethylene oxide are shown in Figures 1 and 2. As can be seen in Figure 1, the initial turbidity, T_{0d}, of the acetylation solution declined linearly for the first 30 minutes of the reaction and then levelled off. When the samples collected at 30 and 40 minutes - at peak and 10 minutes later - were allowed to stand for one day, the turbidity (line labelled 1d in Figure 1) increased 10-15%. The line labelled 1dc in Figure 1 shows the turbidity of the supernatant fluids obtained by centrifuging the 1-day-old acetylation products collected 30 and 40 minutes into the reaction. The difference (ΔT_c) between the turbidity of 1-day-old whole samples (line 1d) and that of supernatant fluids obtained from them (line 1dc) correlates with the filterability of commercially acetylated pulps. When the supernatant fluids were allowed to stand six more days, the turbidity again increased (line labelled 7dc).

Figure 1 also shows the intrinsic viscosity ($[\eta]$) and the yellowness coefficient (Cy) of the supernatant fluids obtained from the 1-day-old samples. Intrinsic viscosity appears to have increased during the last 10 minutes of the reaction, but this finding is probably an artifact since $[\eta]$ normally decreases with time. The yellowness coefficient also increased as the reaction proceeded.

Figure 2 presents the absorbances of the acetylation samples; the lines are labelled like those for turbidity in Figure 1. In general, the absorbance of the sample collected 40 minutes into the reaction was greater than that of the sample taken after 30 minutes, but the range of absorbances is small. Absorbance does not track reaction completeness. The time required to reach 100 units of turbidity (indicated by t_{100} = 28.5 min in Figure 1) is a measure of the speed of the acetylation reaction. The effects of hydroxyethylation, moisture conditioning, and hammering during pretreatment on reaction rate are illustrated in Figure 3. Bone-dry nitration-grade pulp, with or without hammering during pretreatment, requires 114 minutes or more to reach 100 units turbidity. Hydroxyethylation with 2.5% ethylene oxide brings this time down to the region of 50 to 60 minutes for bone-dry pulps; at 7.5% ethylene oxide, the times are 35 to 40 minutes. At all three levels of hydroxyethylation, hammered bone-dry pulps react about 10% more rapidly than pulps that are not hammered during pretreatment. Moisture-conditioned pulps uniformly react more rapidly than bone-dry pulps. Moreover, moisture-conditioned control pulps, with or without hammering during pretreatment, react somewhat faster than hammered bone-dry pulp treated with 7.5% ethylene oxide. Hammered moisture-conditioned control pulp also reacts faster than moisture-conditioned control pulp that is not hammered during pretreatment. In contrast, all hydroxyethylated moisture-conditioned pulps react more rapidly when they are

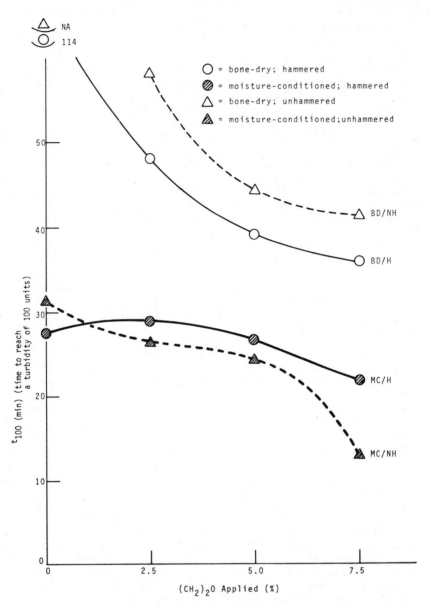

Figure 3. *Effect of hydroxyethylation, moisture conditioning, and hammering on the rate of acetylation of a nitration-grade pulp*

not hammered. The fastest reacting hydroxyethylated pulp -
moisture-conditioned unhammered pulp treated with 7.5% ethylene
oxide - acetylates two to three times faster than the moisture-
conditioned control pulps.

The effects of hydroxyethylation, moisture conditioning, and
hammering on initial turbidity, T_{0d}, of the acetylation product
are illustrated in Figure 4. The moisture-conditioned hammered
control yields a product with an initial turbidity of 105. When
hammering is omitted during pretreatment, T_{0d} increases to 530.
Drying the control pulp further increases initial turbidity to
more than 13,000. The acetylation product of moisture-conditioned
unhammered hydroxyethylated pulp treated with 7.5% ethylene oxide
has the lowest initial turbidity; it is about 20% of that of the
moisture-conditioned hammered control. Hammering improves the
performance of all bone-dry pulps but hurts the performance of
moisture-conditioned hydroxyethylated pulps. Hammered bone-dry
hydroxyethylated pulp treated with 7.5% ethylene oxide results in
an acetylation product with an initial turbidity equal to that of
the moisture-conditioned hammered control.

Filtration characteristics are most strongly associated with
the difference between the turbidity of the whole acetylation
solution and that of the supernatant fluid separated from it by
centrifugation (ΔT_c). The lower ΔT_c, the lower the concentration
of filterable particles. Figure 5 shows ΔT_c for the pulps tested
in this investigation. The control pulp must be hammered and
moisture-conditioned to yield a product with a low ΔT_c. The
particle population, as measured by ΔT_c, is more than two orders
of magnitude higher for the acetylated bone-dry control pulp.
However, the ΔT_c of bone-dry pulp treated with 7.5% ethylene oxide,
with or without hammering during pretreatment, approximates the
ΔT_c of the moisture-conditioned hammered control. When moisture-
conditioned pulp treated with 7.5% ethylene oxide is not subjected
to hammering, its ΔT_c is 75% lower than that of the control.
Hammering the hydroxyethylated pulps is obviously deleterious; for
pulps treated with 5 and 7.5% ethylene oxide, hammering
approximately triples ΔT_c.

The turbidity of the supernatant fluid immediately after
centrifugation, T_{1dc}, is an index of the population of particles
less than 10μ in diameter. Figure 6 shows that moisture conditioning is the major process variable affecting this property.
Although hammering has a large effect on T_{1dc} of the control pulp,
it has a relatively small effect on T_{1dc} of moisture-conditioned
hydroxyethylated pulps. Moisture conditioning combined with
hydroxyethylation greatly reduces the amount of second-phase
materials present in the supernatant fluid. Products from
moisture-conditioned pulps treated with 5.0 and 7.5% ethylene
oxide have T_{1dc} values that are 10% or less of the value for the
moisture-conditioned hammered control. The same relationship
exists when T_{1dc} for products from bone-dry hydroxyethylated pulps
is compared with T_{1dc} for the bone-dry hammered control.

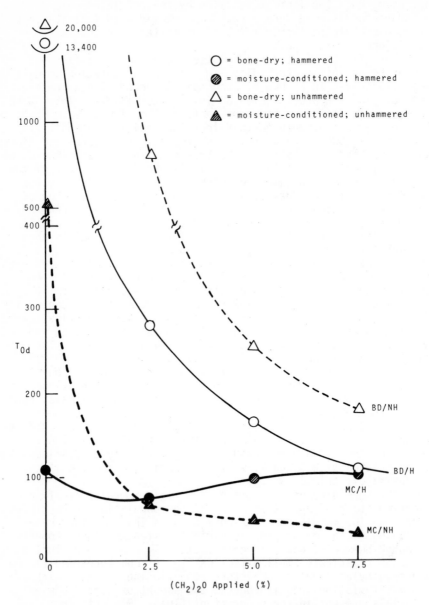

Figure 4. Effect of hydroxyethylation, moisture conditioning, and hammering on the initial turbidity, T_{od}, of acetylation products produced from a nitration-grade pulp (samples collected 40 min into the acetylation reaction)

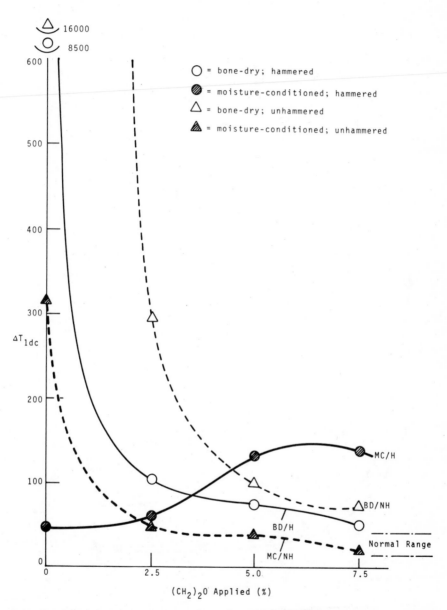

Figure 5.   Effect of hydroxyethylation, moisture conditioning, and hammering on centrifugable particles, ΔT_{1dc}, of acetylation products produced from a nitration-grade pulp (samples collected 40 min into the acetylation reaction)

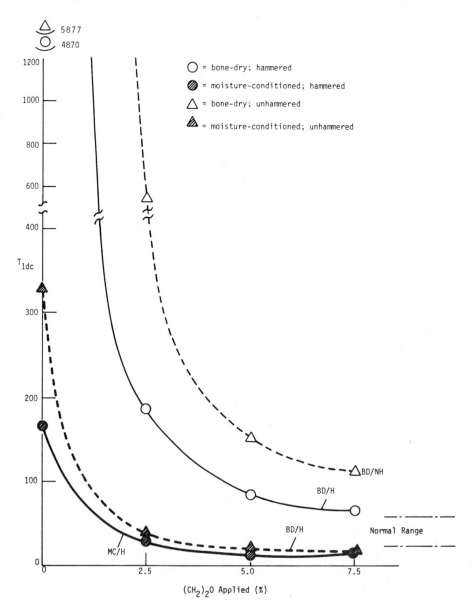

Figure 6.   Effect of hydroxyethylation, moisture conditioning, and hammering on super-nate haze after one day, T_{1dc}, of acetylation products produced from a nitration-grade pulp (samples collected 40 min into the acetylation reaction)

Hammering bone-dry hydroxyethylated pulps is consistently associated with lower supernatant turbidities.

The stability of the small particle population in acetylation products over time correlates with the stability of filtration characteristics; a product poor in this respect will seriously deteriorate after an initial filtration. Moreover, when solutions with unstable small particle populations are extruded, the resultant yarns are unstable with respect to dyeing characteristics.

The turbidity change in an acetylation product on standing, uncentrifuged, for one day plus the turbidity change of the supernatant fluid separated from that product after one day on standing for six more days is used to characterize stability. Figure 7 shows the effects of hydroxyethylation on the stability of the small particle populations in acetylation products of the test pulps. The moisture-conditioned hammered control is far superior to the unhammered or the bone-dry control pulp products. Addition of only 2.5% ethylene oxide results in products equal to or better than those from the optimally treated control. Products of pulps treated with 5.0 and 7.5% ethylene oxide are almost perfectly stable, unless they are moisture-conditioned and hammered. This finding is consistent with the relative disadvantages, mentioned earlier, of hammering moisture-conditioned hydroxyethylated pulps. Most of the difference between hammered moisture-conditioned hydroxyethylated pulps and the other hydroxyethylated pulps develops during the first day of standing (Figure 8). No further appreciable changes occur up to the seventh day (Figure 9).

The relation of turbidity to absorbance enables one to infer something about the nature of particle populations in acetylation products. Turbidity is, of course, the measure of light scattered by suspended particles; absorbance is the measure of the reduction in light transmittance from all causes including light scattering. Because of space limitations, absorbance data corresponding to the turbidity data will not be presented. Instead, the ratio of turbidity to absorbance, T/A, will be discussed.

A transparent particle will scatter as much light as an opaque particle of the same size which differs in refractive index from the medium by the same amount, and therefore will have the same turbidity. However, the absorbance will be higher for the opaque particle. Consequently, T/A will be high for the transparent particle and low for the opaque particles; in this case it will behave as if it is somewhat opaque and its T/A value will be higher. An unacetylated fiber, which has a higher refractive index, will scatter more light and have a higher T/A than a corresponding acetylated, but undissolved, fiber of lower refractive index. A composite particle or agglomerate, characterized by regions of different refractive index, will have a higher absorbance and hence a lower T/A. The more nearly a given particulate mass is mono-disperse at just over half mean λ diameter, the greater the scattering, i.e., the higher the T the higher the concentration of particles, the higher the probability

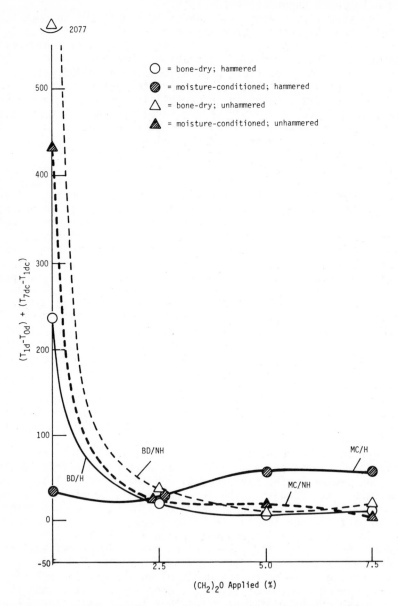

Figure 7. Effect of hydroxyethylation, moisture conditioning, and hammering on overall change in supernate haze in one week, $(T_{1d} - T_{0d}) + (T_{7dc} - T_{1dc})$, *of acetylation products produced from a nitration-grade pulp (samples collected 40 min into the acetylation reaction)*

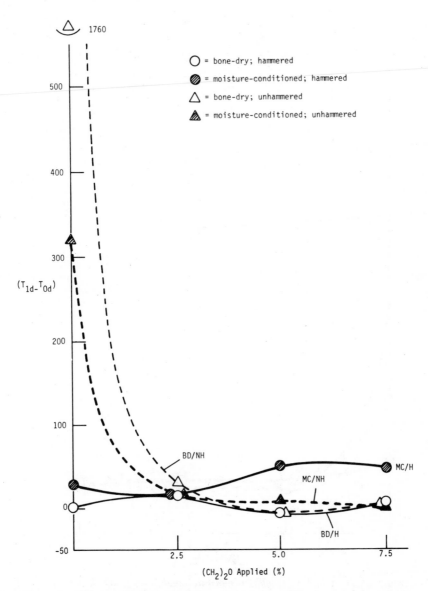

Figure 8. *Effect of hydroxyethylation, moisture conditioning, and hammering on haze change in first day of standing, $(T_{1d} - T_{0d})$, of acetylation products produced from a nitration-grade pulp (samples collected 40 min into the acetylation reaction)*

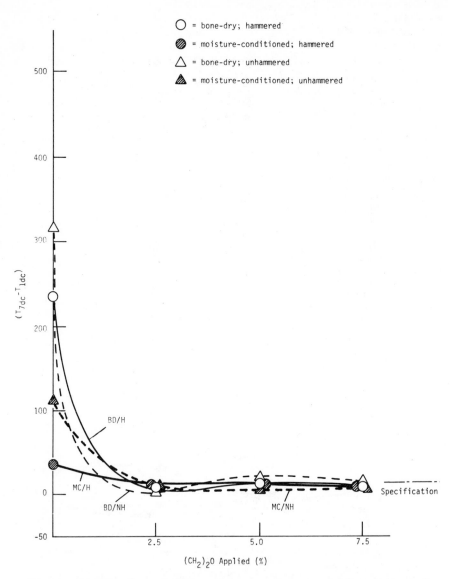

Figure 9. Effect of hydroxyethylation moisture conditioning, and hammering on late haze development in supernate, $(T_{7dc} - T_{1dc})$, of acetylation products produced from a nitration-grade pulp (samples collected 40 min into the acetylation reaction)

of scattered light encountering a particle and resultant
secondary scattering. Scattering in the direction of the beam
will be detected as a "reduction in absorbance"; the result is a
decrease in T/A. Theoretical treatments of scattering are largely
confined to regular shapes, usually spheres, that are homogeneous
in refractive index. Interpretation of data is not to be
approached with simplistic models. The relatively low concentration of particles in most samples after thirty minutes of
acetylation minimizes coincidence effects, such as shadowing of
one particle by another, and secondary scattering.

With these concepts in mind, it is interesting to see what
effects moisture conditioning, hammering, and hydroxyethylation
have on the particle population of acetylation products. For the
control pulp, the ratio of turbidity to absorbance for the
materials removed by centrifugation, $\Delta T_c/\Delta A_c$, is significantly
higher under bone-dry conditions that it is after moisture conditioning (Figure 10) because fibers are present in greater
amounts (Figure 4) and they have a lower degree of acetylation.
Hammering has no effect on $\Delta T_c/\Delta A_c$ for the moisture-conditioned
control pulp product and only a small effect on the ratio for
bone-dry control pulp. The lowest $\Delta T_c/\Delta A_c$ values are seen with
hydroxyethylated pulps produced by adding 5.0 and 7.5% ethylene
oxide and using moisture conditioning but no hammering. Hammering is particularly deleterious for pulps treated with 7.5%
ethylene oxide. In contrast, for the bone-dry hydroxyethylated
pulps treated with 2.5 and 5.0% ethylene oxide, hammering is
beneficial. The most striking effect of process conditions on
$\Delta T_c/\Delta A_c$ is observed for the unhammered hydroxyethylated pulp
treated with 7.5% ethylene oxide; moisture conditioning causes a
fivefold decrease in T/A for this pulp.

The nature of the particles that remain suspended in the
acetylation product after centrifugation is different from that
of the particles that settle out (Figure 11). Bone-dry pulps
yield supernatant fluids with the highest T/A values, but only
for the control pulp does T/A exceed 2.2. For centrifugable
particles, $\Delta T_c/\Delta A_c$ exceeded 2.2 in 13 out of 16 cases (Figure 10).
The moisture-conditioned pulps yield supernatant fluids with the
lowest T/A values. For hydroxyethylated moistured-conditioned
pulps the range is 0.3 to 0.7. The hammered moisture-conditioned
hydroxyethylated pulps have T/A values about 0.25 lower than
their unhammered counterparts.

The nature of the particles formed on standing is evaluated
by subtracting the T/A ratio for the 1-day-old supernatant fluid
from that for the 7-day-old supernatant fluid (Figure 12). For
the hydroxyethylated pulps, the materials formed on standing
generally have a slightly higher T/A ratio than the particles
initially present. This is not the case for the control pulps.
Although standing has little or no effect on T/A for the
moisture-conditioned control samples, appreciable differences
are observed for the bone-dry control samples. Late formed

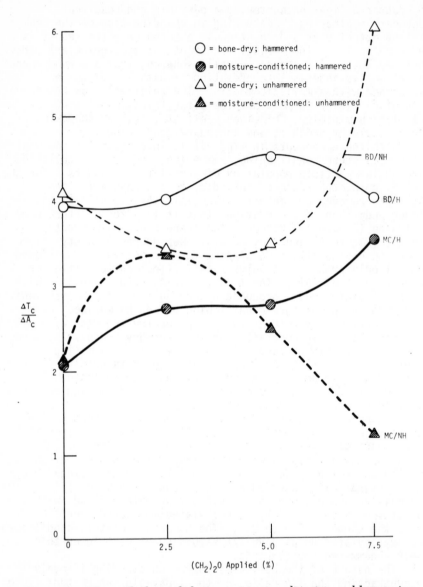

Figure 10. Effect of hydroxyethylation, moisture conditioning, and hammering on optical character, $\Delta T_c/\Delta A_c$, of acetylation products produced from a nitration-grade pulp (samples collected 40 min into the acetylation reaction)

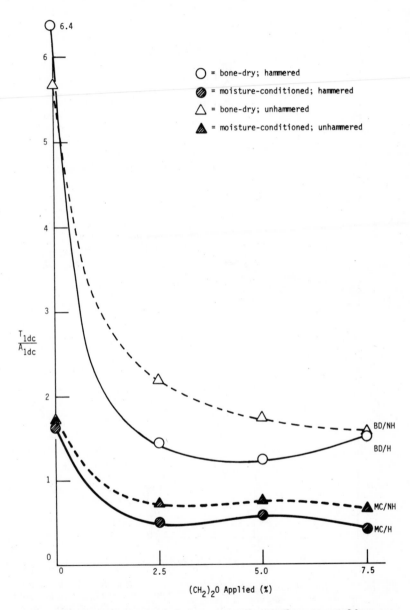

Figure 11. Effect of hydroxyethylation, moisture conditioning, and hammering on optical character of haze, T_{1dc}/A_{1dc}, of acetylation products produced from a nitration-grade pulp (samples collected 40 min into the acetylation reaction)

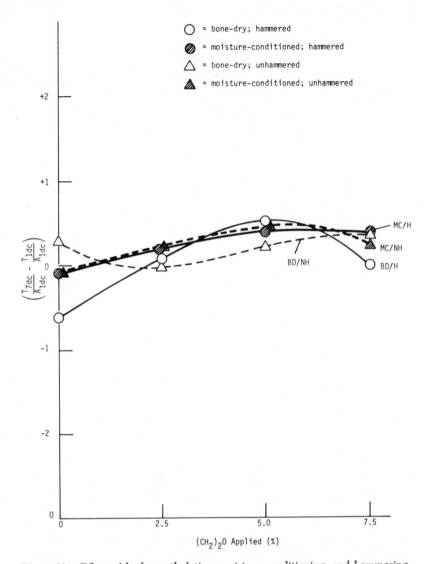

Figure 12. Effect of hydroxyethylation, moisture conditioning, and hammering on optical character of late haze [$(T_{7dc}/A_{7dc}) - (T_{1dc}/A_{1dc})$] of acetylation products produced from a nitration-grade pulp (samples collected 40 min into the acetylation reaction)

particles have a lower T/A value for hammered bone-dry pulps and a higher value when hammering is not used.

As previously mentioned, a high T/A ratio indicates a particle of relatively high refractive index. The refractive index of unacetylated cellulose is 1.54; that of cellulose acetate is 1.47. For the medium, which is predominantly acetic acid, it is 1.37. The centrifugable particles, which have a T/A value of 2 to 4, are high in unacetylated cellulose. The small disperse particles in the supernatant fluid, which have a T/A value of 0.5 to 2, are acetylated and probably partly solvated. The increase in the T/A value of the supernatant fluid on standing may be due to a decrease in solvation; such a change would result in particles with a higher optical density.

Based on the high T/A value of 6 (Figure 10), the centri-fugable particles from unhammered bone-dry hydroxyethylated pulps treated with 7.5% ethylene oxide appear to be highly cellulosic. They are probably aggregations of crystalline cellulose micro-fibrils with a minimal matrix of esterified cellulose. Figure 13 is an electron micrograph of similar residues obtained from an acetylated dissolving-grade pulp. The low $\Delta T_c/\Delta A_c$ found for the centrifugable particles from moisture-conditioned unhammered hydroxyethylated pulp treated with 7.5% ethylene oxide suggests that these particles are mostly esterified materials with sufficient interconnections between elements to provide a struc-tural integrity which is not susceptible to acetic acid. Cellulose triacetate crystallites are not soluble in acetic acid and may act as tie points. Cross-links involving the hydroxyethyl group are also a possibility.

The intrinsic viscosity of the acetylation product 30 minutes into the reaction is independent of the degree of hydroxyethy-lation for moisture-conditioned unhammered pulps (Figure 14). Hammering reduces the apparent intrinsic viscosity* as the degree of hydroxyethylation increases. The acetylation of bone-dry control pulp and of bone-dry pulp treated with 2.5% ethylene oxide had not progressed sufficiently in 30 minutes to provide solutions suitable for intrinsic viscosity measurement. For bone-dry pulps with normal reactivity, such as one treated with 5% ethylene oxide, the intrinsic viscosity is comparable to that of moisture-condit-ioned samples. The higher intrinsic viscosity of pulps treated with 7.5% ethylene oxide suggests that in the bone-dry system there may be an appreciable amount of cross-linking between the hydroxyl group of the hydroxyethyl substituent and uronic acids. In the presence of acetic anhydride and a catalyst, the mixed uronic-acetic acid ester forms rapidly, making this a chemical possibility.

The yellowness of acetylation solutions is primarily due to

---

*Intrinsic viscosities were obtained by conversion of product viscosities; it was not determined whether the conversion factor differed for hydroxyethylated pulps.

Figure 13.   Transmission electron micrograph of film made from an acetylation mixture.
Reference mark is 1μ.

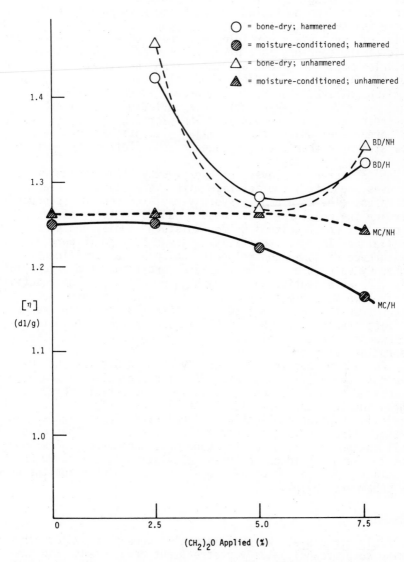

Figure 14.   Effect of hydroxyethylation, moisture conditioning, and hammer-ing on intrinsic viscosity of acetylation products produced from a nitration-grade pulp (samples collected 30 min into the acetylation reaction)

the selective scattering of blue light by particles approximately half the wavelength of blue light in diameter. A measure of the formation of such particles in the process is therefore provided by the yellowness coefficient, Cy (Figure 15). The highest population of these small particles was measured in the supernatant fluid of the bone-dry control pulp. The T/A value of these particles was high (about 6), suggesting that the particles are probably fragments of unacetylated cellulose, most likely microfibril bundles 200-250 mμ in diameter (Figure 13). These fragments separate as a result of the acetylation and solvation of the low-order hemicellulose matrix. The microfibril bundles are dense and unreactive because the fibrillar elements have not been separated by either the water/acetic acid pretreating medium or the hydroxyethyl wedges.

With increasing levels of hydroxyethylation, reactivity improves and actual contact time with the acetylation reagent increases. The more highly hydroxyethylated bone-dry products approach the moisture-conditioned products in terms of Cy. The generally lower levels of Cy for moisture-conditioned samples parallel the improved reactivities shown by t_{100} (Figure 3). The moisture-conditioned control pulp, which has a significantly higher Cy value than do the moisture-conditioned hydroxyethylated pulps, is an exception in having high reactivity. The reactivity retained in microfibriller bundles of hydroxyethylated pulps is higher than that which moisture conditioning alone can induce. The base level of Cy, about 0.07, indicates the contribution to yellowness of extractable components which have a molecular absorption function.

As can be seen in Figure 16, Cy changed between the 30- and 40-minute acetylation samples. With the exception of the bone-dry, unhammered pulps, this change involved substantial increases, suggesting that the particles scattering blue light at 30 minutes remain in solution and that additional particles are generated by further attack on large, centrifugable particles. The relatively small changes observed for the bone-dry, unhammered pulps suggests that the large residues in those acetylations are less vulnerable to attack. The values of ΔT_c were also much higher for the bone-dry unhammered pulps.

Discussion

Pulp Structure. By what mechanism does pulp hydroxyethylation so profoundly modify pretreatment requirements and the acetylation process product? The formulation evolved from the findings of this investigation is based on the assumption that the ultimate element in the cellulosic structure is a microfibril 35 Å in diameter. Electron micrographs of the residues after peak acetylation regularly show composite rod bundles whose diameters are multiples of 35 Å. Also present are particles in which an amorphous cementing agent apparently holds together dense, roughly

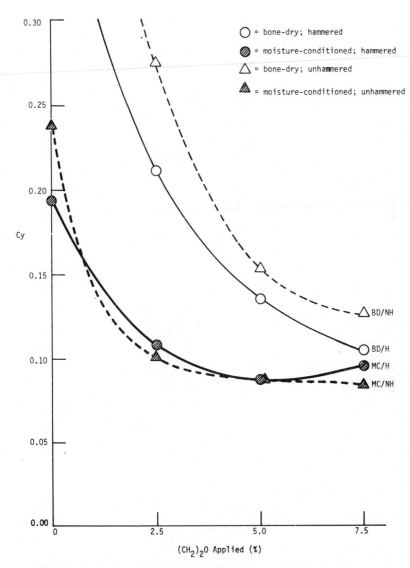

Figure 15.   Effect of hydroxyethylation, moisture conditioning, and hammering on yellowness coefficient, Cy, of acetylation products produced from a nitration-grade pulp (samples collected 30 min into the acetylation reaction)

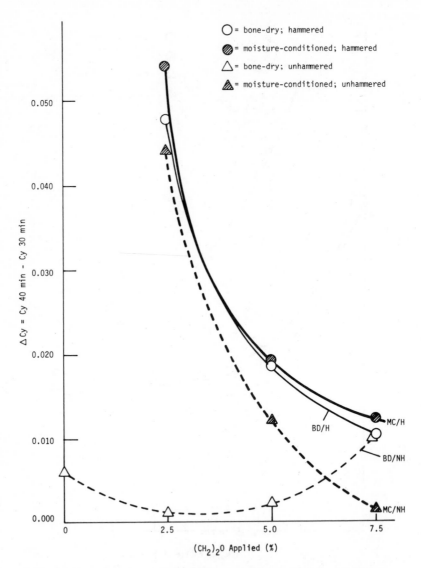

Figure 16. Effect of hydroxyethylation, moisture conditioning, and hammering on the change in the yellowness coefficient, ΔCy, as the acetylation of a nitration-grade pulp progresses (samples collected 30 and 40 min into the acetylation reaction)

oval-shaped particles (Figures 13 and 17). These particles are hemicellulosic, in that they are not observed in acetylated cotton fibers.

Pretreatment

The optimum pretreating liquid is a water/acetic acid solution at a 1/1 molar ratio. In the absence of water acetic acid is a dimer, but in the presence of water the dimer decomposes. The actual pretreating agent is probably the mono-hydrate of acetic acid. It is bulkier than H_2O and therefore a better wedge. It is, unlike dimeric acetic acid, a hydrogen donor. It can be postulated that the monohydrate of acetic acid is effective in breaking hydrogen bonds in delignified cellulosic material and therefore that it effects swelling and opening of the larger microfibrillar units, thereby making available a larger surface for the attack of the acetylation reagent. Mechanical input (hammering) aids the penetration of the acetic acid monohydrate. Other studies have shown that the reactivity of a pulp with an extremely low hemicellulose content is not improved by mechanical input during pretreatment. However, where hemicellulosic cements are present, mechanical input breaks the bonds of these cements and creates better access for the acetic acid monohydrate.

Hydroxyethylation

When undried pulp at a consistency of 100% is treated with ethylene oxide, the accessible surface area is at least as great as that resulting from the pretreatment of dried pulp. Ethylene oxide enters the interfibrillar spaces, randomly reacts at a number of accessible sites, and provides a series of wedges which prevent structural collapse as drying takes place. When the degree of hydroxyethylation is sufficient, a bone-dry hydroxyethylated pulp, which has as open an interfibrillar structure as a pretreated pulp, will react at a rate similar to that characteristic of moisture-conditioned pulps. Supporting evidence for this theory can be found in the significantly lower number of large residues, as indicated by the lower ΔT_c values, in hydroxyethylated pulps. Hammering a bone-dry hydroxyethylated pulp generates even more surface area and results residual particles.

Moisture conditioning plasticizes hydroxyethylated pulp. Hammering these plasticized pulps leads to adhesions where hydrogen bonds develop in regions which were separated but in which the wedge incidence was small. The result is an increase in large, centrifugable particles, indicated by the high initial turbidity and the high ΔT_c values of moisture-conditioned hammered hydroxyethylated pulps.

The wedge effect of hydroxyethylation in conjunction with the acetic acid monohydrate pretreatment also results in a reduction of particles in the size range of 0.5 to 10μ. The basis

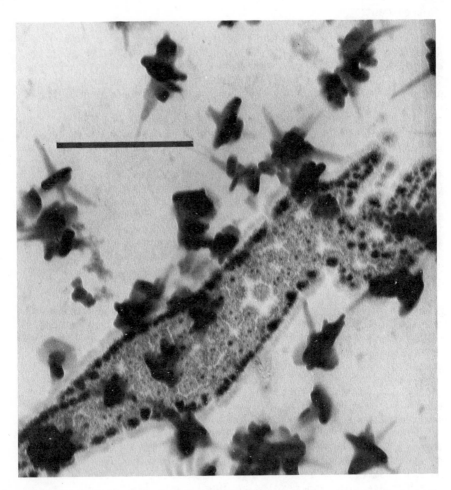

*Figure 17.   Transmission electron micrograph of film made from an acetylation mixture.
Reference mark is 0.1μ.*

for this observation is the low turbidities of supernatant fluids
from hydroxyethylated pulps. This mechanism is compatible with
the observation of a higher cellulose content in particles formed
when bone-dry pulps are acetylated.

Conclusions

All three levels of ethylene oxide treatment of a nitration
grade pulp lead to acetylation products with improved filtration
performance, reduced false viscosity, and increased acetone
solution clarity. Most strikingly, the hydroxylated pulps could
be acetylated bone-dry without activation by mechanical input
(hammering) during pretreatment.

The water solubility of the hemicellulose constituents of the
hydroxyethylated pulps was increased. As cellulose acetate is
usually precipitated in dilute aqueous acetic acid and washed free
of acid with water, this increased water solubility of contaminants results in serious yield reductions in the precipitation and
washing stages.

Although improvement in acetylation products produced from a
lower quality pulp was achieved by hydroxyethylation, the economic
and technical limitations of this approach suggest that treating
less purified pulps with ethylene oxide is not a viable commercial
process and does not solve the problem at hand. In the context of
basic scientific knowledge, however, the findings of this investigation are considerable.

In terms of a mechanism, a structurally modified lower grade
pulp acetylates rapidly and forms a better product because much of
the initial before-drying internal surface is preserved by hydroxyethyl groups which act as molecular wedges, preventing
extensive hydrogen bonding and attendant reduction in surface.
The customary pretreatment with acetic acid monohydrate and
moderate mechanical treatment improve the acetylation response of
the parent pulp by regenerating internal surface. In contrast,
moderate mechanical treatment of an acetic monohydrate pretreated
hydroxethylated pulp reduced internal surface by crushing the
structure and establishing hydrogen bonds, thereby reducing
internal surface and accessibility to acetylation reagents.

11

Reactions of Cellulose With Amic Acids and Anhydride-Ammonia

Part VI: Reaction Stoichiometry

B. G. BOWMAN* and JOHN A. CUCULO

Dept. of Textile Chemistry, North Carolina State University, Raleigh, N.C. 27607

The chemical and physical properties of cellulosic fabrics can be modified in a pad-bake reaction with solutions of certain half-amides of dicarboxylic acids (amic acids) (1, 2, 3). The products of the reaction are ammonia and cellulose half-acid esters (Equation 1). Woven and nonwoven fabrics composed of

$$CellOH + R \begin{array}{c} \overset{O}{\overset{\|}{C}} - OH \\ \\ \underset{O}{\overset{\|}{C}} - NH_2 \end{array} \xrightarrow{\Delta} CellO\overset{O}{\overset{\|}{C}}R\overset{O}{\overset{\|}{C}}-OH + NH_3 \qquad (1)$$

either viscose rayon or cotton (1, 2, 3) and wood pulp mats (4) have been modified by this procedure. The reaction is specific for amic acids derived from dicarboxylic acids which form internal pentacyclic anhydrides (e.g., succinic, maleic, phthalic acids, etc.) (3). The reaction also occurs when solutions of anhydrides and aqueous ammonia are reacted with the cellulosic fabrics (Equation 2) (5).

The reaction is moderately catalyzed by small amounts of a variety of compounds and may be successfully completed when the pad-bath solvent is either water or formamide (4). The reaction does not appear to be successful when the pad-bath solvent is an aprotic medium such as N,N-dimethylformamide or N-methylpyrrolidone (4).

*Present Address: High Point College, Department of Chemistry, High Point, N. C. 27262

$$\text{(structure)} + NH_3 \text{ (aq.)} \longrightarrow \text{(structure)} + CellOH \xrightarrow{\Delta}$$

(2)

in situ

$$CellO\overset{O}{\overset{\|}{C}}- \qquad \overset{O}{\overset{\|}{C}}OH + NH_3$$
$$\diagdown_{C \ - \ C} \diagup$$

The pad-bake process involves the application of the amic acid and catalyst in solution to the fabric, squeezing to a predetermined pick-up, and placing in a hot oven to remove the solvent and to effect the reaction. The solvent is rapidly removed as vapor, typically in about two minutes, leaving behind a solid or liquid residue on the fabric. The rapid removal of the solvent, water, has been shown to be important (4)

The fabric modified by the pad-bake reaction with α, β -amic acids has increased base combining capacity and water sensitivity. This indicates that acid groups are indeed present on the fabric. The saponification equivalent of the fabric is generally twice the neutralization equivalent indicating the presence of equal numbers of acid and ester groups. Longer reaction times or higher temperatures give rise to the formation of diester crosslinks (Equation 3) (1 ,4). Under these conditions the saponification equivalent is more than twice the neutralization equivalent. There is some evidence that a small amount of half-amide

$$CellO\overset{O}{\overset{\|}{C}}CH_2CH_2\overset{O}{\overset{\|}{C}}OH + CellOH \longrightarrow CellO\overset{O}{\overset{\|}{C}}CH_2CH_2\overset{O}{\overset{\|}{C}}OCell$$

(3)

ester is formed (3, 4). Spectroscopic evidence for the formation of the half-acid ester of cellulose by the pad-bake reaction with α, β-amic acids is given in Figure 1. These infrared spectra were obtained from cupraphane film which was modified in the pad-bake reaction with succinamic acid. Spectrum A represents infrared absorption of the unmodified cellulose in the carbonyl absorption region. There is one weak doublet near 1650 cm^{-1} which is due to the presence of water. The carbonyl absorption region of the cellulose hemisuccinate is shown in spectrum B. The strong absorption at 1730 cm^{-1} is due to the overlapping carboxylic acid and ester carbonyl absorptions. The acid-ester combination peak is resolved in spectrum C

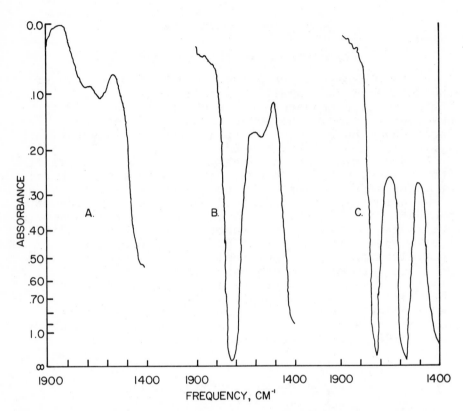

Figure 1.   Carbonyl absorption region for (A) unmodified cellulose, (B) cellulose hemi-succinate, and (C) cellulose hemisuccinate, sodium salt

when the cellulose hemisuccinate is neutralized with
sodium bicarbonate. The peak at 1730 cm^{-1} decreases in
intensity and increases in sharpness as a new peak due
to carboxylate appears at 1570 cm^{-1}. These assignments
are consistent with the literature (6, 7).

The reaction between the α, β-amic acids and
cellulose is significant because the derivative, the
cellulose half-acid ester, can be rapidly formed using
relatively inexpensive starting materials and aqueous
systems. This is not generally true for the other
methods of preparation of cellulose half-acid esters
which have been reported (8).

II. Discussion of the Probable Mechanisms of the Reaction Between Cellulose and α, β-Amic Acids

The reaction of cellulose with α,β-amic acids
represents an apparent departure from the traditional
observations of amide chemistry. The reaction is
essentially an alcoholysis of the amide group as the
overall stoichiometry indicates (Equation 1). The
amide group is a relatively selective acyl moiety and
requires a strong nucleophile for attack at the
trigonal carbon atom of the group. The attack of an
alcohol, a weak nucleophile, on the amide group with
subsequent loss of ammonia is not a very favorable
reaction. Indeed, March states in Advanced Organic
Chemistry: Reactions, Mechanisms and Structure that
"Alcoholysis of amides is possible but is seldom
performed . . ." (9). The reason for the inertness of
amides is the high basicity of the amide ion which is
the leaving group expected in the uncatalyzed
alcoholysis (Equation 4).

$$R-\overset{\overset{\displaystyle O}{\|}}{C}\diagdown_{NH_2} + R'OH \longrightarrow R-\overset{\overset{\displaystyle |\overset{\ominus}{O}|}{|}}{\underset{NH_2}{C}}-\overset{\oplus}{O}\diagup^{H}_{\diagdown R'} \longrightarrow R\diagup^{\overset{\displaystyle O}{\|}}{C}\diagdown_{\overset{\oplus}{O}}\diagup^{H}_{\diagdown R'} + \overset{\ominus}{NH_2} \longrightarrow$$

$$R\diagup^{\overset{\displaystyle O}{\|}}{C}\diagdown_{O}\diagup^{R'} + NH_3 \qquad\qquad\qquad (4)$$

This is the same reasoning used to explain the relative
inertness of amides toward neutral hydrolysis (9).

The probable mechanism by which α,β-amic acids react with cellulose can be divided into two general classes:

1. Direct alcoholysis by attack of the cellulosic hydroxyls at the amide carbonyl; and
2. Alcoholysis of a reactive intermediate formed by the initial decomposition of the amic acid.

The direct alcoholysis may be either catalyzed by acid or uncatalyzed. The uncatalyzed reaction mechanism is shown in Equation 4. The mechanism proposed by Johnson and Cuculo involves intermolecular acid catalysis(Equation 5) (2). Attack by a cellulosic hydroxyl on the amide is assisted by the initial protonation of the amide and by the formation of

(5)

a hydrogen bond between the alcoholic proton and the carboxyl carbonyl. The justifications given for this reaction mechanism were the apparent specificity of the reaction for α, β-amic acids derived from dicarboxylic

acids which form internal pentacyclic anhydrides and
the catalytic effect of acids. Another source of
catalyzing acid is the adjacent carboxyl group. This
leads to the possibility of intramolecular acid
catalysis as shown in Equation 6.

The second classification of mechanisms involves
the formation of reactive intermediates which can more
easily undergo alcoholysis. Since α, β-amic-acids are
characteristically easy to hydrolyze (10, 11, 12, 13,
14, 15, 16, 17, 18) there is good reason to suspect
that in an aqueous pad-bath, the amic acid could be
hydrolyzed. This hydrolysis reaction would yield both
the diacid and ammonia which could subsequently be
deposited as the partial

$$\tag{6}$$

ammonium salt when the water is removed. Esterification of cellulose by ammonium salts has been reported
(Equation 7) (19, 20, 21).

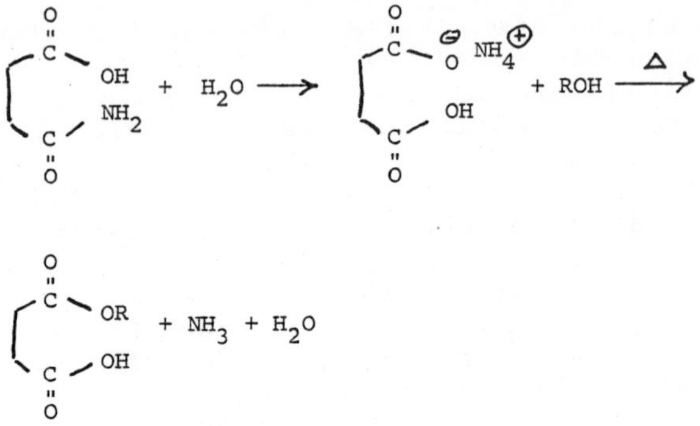

$$\text{(7)}$$

While this reaction could contribute to the formation
of the cellulose half-acid ester, the experimental
results of Cuculo and Johnson (2) have indicated that
the ammonium salt reaction is not the major contributor
in reactions of succinamic acid with viscose rayon.
    An alternative reactive intermediate is the
anhydride which has been proposed as the intermediate
in the hydrolysis (10, 11, 12, 13, 14, 15, 16, 17, 18)
and aminolysis (22) reactions of α,β -amic acids.
Anhydride might form in the residue left on the fabric
after the pad-bath solvent is removed and subsequently
react with the cellulose (Equation 8). This mechanism
is

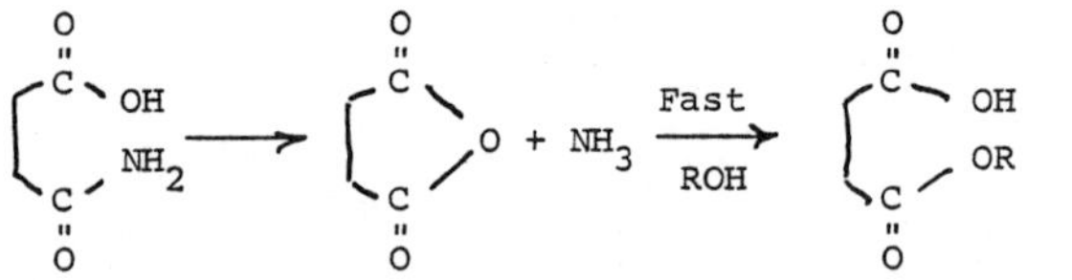

$$\text{(8)}$$

attractive for the following reasons:

1. Formation of anhydride from α, β-amic acids has
    been directly detected in several cases (18,
    23, 24, 25) and
2. The anhydride in much less selective than the
    amide in solvolysis reactions and easily under
    goes alcoholysis at room tempeature (26).

    This paper describes a study undertaken to
determine the chemical changes which occur in
succinamic and phthalamic acids during the preparation
of the pad-bath and the pad-bake reaction. The

pad-bake procedure was simulated under conditions which allowed complete recovery of the residue of reaction. The reactions were performed in the absence of cellulosic substrate in the hope that any stable intermediate which might be the precursor to the cellulose half-acid ester might be directly detected. Also sought were the products of reactions in competition with the esterification. The composition of the residue was studied as a function of bake time and temperature, initial concentration of the bath, presence of catalyst, solvent, and drying conditions (removal of solvent).

Experimental

Preparation of Amic Acid Solutions. Phthalamic acid or succinamic acid was prepared by the methods given in references (4) and (27). The amic acid solution was prepared in a 10 ml volumetric flask maintained at $60 \pm 3\,°C$ in a water bath. The distilled water used to prepare the amic acid solution was incubated in the water bath. When catalyst was used it was added after the amic acid.

Performing the Reaction.   An aliquot of amic acid solution was removed from the flask with a preheated constant delivery syringe and injected through a rubber septum into a preheated 25 ml reaction flask. The flask was heated by a large volume oil bath held at the reaction temperature $\pm 1\,°C$. The initial amount of amic acid was determined from the solution concentration and aliquot volume. The reaction flask was fitted with a nitrogen inlet and condenser. Nitrogen at a slight positive pressure flowed through the reaction flask and out through the condenser into a flask containing standard acid to trap any volatile ammonia. In some cases the acid trap was preceded by a chloroform trap. The duration of the reaction was taken as the time from injection of the solution until the removal of the reaction flask from the oil bath for quenching in a stream of acetone. The flask was then cleaned, and dried under vacuum at 40°C for one hour, and its mass determined. Reactions from N, N-dimethylformamide and formamide required longer periods of drying. The residue mass was determined by difference. The conditions under which the reactions were performed are summarized in Table I.

Analysis of the Products. The reaction products were analyzed qualitatively by infrared spectroscopy. The spectra were determined as KBr pellets using a

Table I

Summary of Reaction Conditions

Reaction Number	Amic Acid	Concentration, molar	Duration of Reaction, min.	Reaction Temperature °C	Bath Solvent
1	s	2.5	10	150	water
2	=	=	=	175	=
3	=	=	=	200	=
4	p	=	=	150	=
5	=	=	=	175	=
6	=	=	=	200	=
7	s	=	5	175	=
8	=	=	15	=	=
9	p	=	5	=	=
10	=	=	15	=	=
11	s	1.5	10	=	=
12	=	4.0	=	=	=
13	p	1.5	=	=	=
14*	s	2.5	=	=	=
15*	p	=	=	=	=
16	s	=	=	=	formamide
17	s	2.5	10	175	DMF
18	p	=	=	=	formamide

Table 1 continued

Reaction Number	Amic Acid	Concentration, molar	Duration of Reaction, min.	Reaction Temperature °C	Bath Solvent
19	p	"	"	"	DMF
20**	s	"	"	"	water
21**	p	"	"	"	"
22	s	melt	"	"	none
23	p	"	"	"	"

s, succinamic acid   p, phthalamic acid   DMF, N,N-dimethylformamide

* 6% ammonium sulfamate based on the no. of moles of the amic acid

** solution evaporated in a stream of compressed air for 12 hrs. before being heated

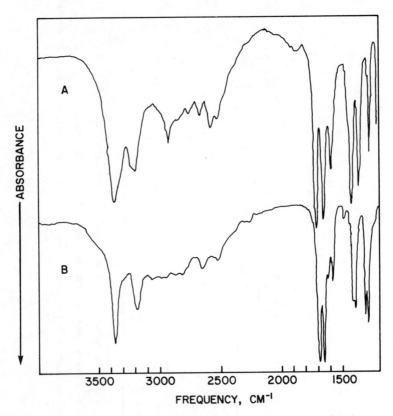

Figure 2. Infrared spectra of (A) succinamic acid and (B) phthalamic acid in the region 4000–1200 cm⁻¹

Perkin-Elmer Infracord 337 Recording Spectrophotometer.
The position of absorption maxima reported here were
measured from polystyrene reference film.

The products were analyzed for ammonium ion by
Nesslerization (28) using standards prepared from
ammonium sulfate.

The neutralization equivalent of the reaction
product was determined by potentiometric titration with
standard ammonium hydroxide. The endpoints were
determined analytically from values of the second
derivative curve in the region of the endpoint. The
analysis with the weak base was necessary because a
portion of the residue was an ammonium salt. The
determination of free acidity in the presence of a salt
of the acid with a weak base is discussed in reference
(29).

The amount of chloroform soluble material in the
reaction product was determined by extraction of a
powdered sample of product with an aliquot of
chloroform. The extraction was performed in a fritted
funnel. The extraction was considered complete when
the decrease in mass of the residue on successive
extractions was less than 0.1% of the initial mass.
The chloroform extracts were recovered by evaporation
and their infrared spectra and melting ranges were
determined.

Total nitrogen in the samples was determined by
the micro-Kjeldahl technique (30).

Determination of the Stability of the Amic Acid
Solutions. The stability of the solutions prepared as
above was determined by following the appearance of
ammonium ion with time and by examination of the
infrared spectrum of the material in the solution. In
both analyses an aliquot of solution was withdrawn and
Nesslerized for ammonium content or evaporated to
dryness under vacuum for evaluation of the infrared
spectrum as described above.

Results of the Experiments

The infrared spectra of succinamic and phthalamic
acids are given in Figure 2. The distinctive features
of the amic acid spectra are the carbonyl doublet near
1700 cm^{-1} and the doublet in the region 3000 cm^{-1} to
3500 cm^{-1} due to the nitrogen-hydrogen stretching
vibrations. The carbonyl peaks in succinamic acid are
at 1710 cm^{-1} and 1650 cm^{-1}. In phthalamic acid the
carbonyl peaks occur at 1695 cm^{-1} and 1645 cm^{-1}. The
high energy peak is due to the carboxylic acid carbonyl
stretching vibration and the low energy peak due to the

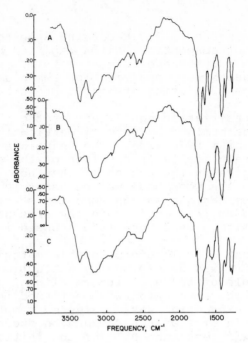

Figure 3. Infrared spectra of the products of the reaction of succinamic acid for 10 min at 150°C (A), 175°C (B), and 200°C (C)

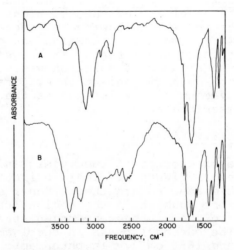

Figure 4.   Infrared spectra of (A) succinimide and (B) succinamic acid containing 20 wt % succinimide

amide carbonyl stretching vibration. The nitrogen-
hydrogen stretching vibration peaks occur at 3375 cm^{-1}
and 3215 cm^{-1} in succinamic acid and at 3355 cm^{-1} and
3155 cm^{-1} in phthalamic acid. This doublet corresponds
to the asymmetric and symmetric N-H stretching
vibrations. These assignments are consistent with the
literature (31, 2, 32).

The spectra of the products of the reaction of
succinamic acid for ten minutes at 150°C, 175°C, and
200°C (reactions 1,2, and 3, Table I) are shown in
Figure 3. Comparison of these spectra with that of
succinamic acid in Figure 2 shows that significant
changes have occurred in the nitrogen-hydrogen and
carbonyl stretching regions. In spectrum A, the N-H
stretching peaks are of almost equal intensity, whereas
in the pure amic acid the high energy peak is by far
the most intense. In spectra B and C (175°C and
200°C, respectively) the intensity of the low energy
N-H stretching vibration is the greatest. The amide
carbonyl absorption which has begun to decrease in
spectrum A has become a shoulder in spectra B and C.
This indicates a progressive decrease in the amount of
amide in the residue as the temperature of reaction
increases. In addition the peak which appears near
1575 cm^{-1} in the pure amic acid (probably the amide
II band (6, p.94,95) has broadened and the position of
the maximum has shifted to 1550 cm^{-1} in spectra B and C.
This is an indication of the formation of an ammonium
salt in the residue since the carbonyl stretching
absorption of the carboxylate anion occurs near 1550
cm^{-1} (6). This conclusion is supported by the
appearance of two broad peaks of medium intensity near
2500 cm^{-1} and 2000 cm^{-1} which would be expected in the
spectra of ammonium salts (6, p.94). A further change
which occurs in the carbonyl region is the appearance of
a shoulder at 1760 cm^{-1} in spectrum B which is sharply
resolved in spectrum C. The appearance of the peak at
1760 cm^{-1} can best be explained by the presence of
succinimide in the product. The spectrum of the pure
imide is shown in Figure 4 (spectrum A) along with that
of succinamic acid containing 20% succinimide by weight
prepared from the pure compounds (spectrum B). The
spectra clearly show that the absorption peak at 1760
cm^{-1} can be explained by the presence of succinimide.
Thus it can be concluded on the basis of the infrared
spectra that the relation products include unreacted
succinamic acid, ammonium salt which could be a mixture
of ammonium succinamate, monoammonium succinate, and
diammonium succinate, and succinimide.

The qualitative analysis of the reaction products

by infrared spectroscopy is supported by the
quantitative analytical data presented in Table II
(Reactions 1,2,3). The presence of ammonium ion as
determined by Nessler's reagent indicates that a
substantial amount of hydrolysis has occurred either in
the preparation of the solution or in the bake step.
The amount of ammonium ion is markedly lower at the
highest temperature, 200°C.

The decrease in the amount of ammonium ion with
temperature is accompanied by an increase in the amount
of chloroform soluble material and neutralization
equivalent. This reflects an increasing amount of
succinimide in the residue. The imide is chloroform
soluble and is too weak an acid to be determined by the
neutralization equivalent determination. When imide is
removed from the residue by extraction the residue
contains the acids and salts. The total amount of
nitrogen (as ammonia) present in this residue was
determined by micro-Kjeldahl analysis. This value
represents the total amide and ammonium nitrogen on an
imide-free basis. When based on the total mass of the
residue this value can be used to determine the % amide
ammonia in the mixture. These derived values for % amid
ammonia are given in Table II. The amount of amide
ammonia decreases from 9.19% at 150°C to 7.62% at 200°
C.

By using the values for % chloroform soluble
material, it is possible to calculate the % yield of
imide in the reaction. It is also possible to
calculate the yields of ammonium salt (as monoammonium
succinate) and amic acid from the % ionic ammonia and %
amide ammonia values respectively. It is also possible
to calculate the yield of diacid from the values of the
neutralization equivalent and the yields of monoammonium
salt and amic acid. This last calculation is based on
the assumption that the acids present in the mixture are
the amic acid, the monoammonium salt and the diacid.
The results of such calculations are shown in Table
III. The data for reactions 1, 2 and 3 show that the
amide content decreases while the yields of imide and
diacid increase (as the temperature is increased). The
ammonium salt content decreases. These data are
consistent with the conclusion that hydrolysis occurs in
the preparation of the amic acid solution and during the
volatilization of the solvent. The dehydrocyclization
reaction which yields imide is a competing reaction
which yields an inert product (3).

The spectra of the products of the pad-bake
reaction of phthalamic acid for 10 minutes at 150 (A),
175 (B), and 200°C (C) (reaction 4, 5, and 6, Table I)

Table II

Analytical Data for Reaction Stoichiometry Study

Reaction Number	% Mass Recovered	% CHCl$_3$ Extracts	% Ionic NH$_3$	Neutralization Equivalent	% NH$_3$ in Extract Residue	% Amide NH$_3$
1	102.3	0.43	4.09	115.40	13.33	9.19
2	100.8	5.84	5.27	123.13	13.38	7.35
3	92.4	41.11	0.99	173.11	14.63	7.62
4	107.5	6.95	8.36	196.67	9.32	0.31
5	104.9	12.41	8.08	204.41	9.34	0.00
6	94.9	45.70	5.16	331.85	9.32	0.00
7	99.6	0.62	3.63	117.53	13.33	9.61
8	97.6	21.42	2.78	134.67	11.63	6.35
9	110.7	2.05	8.25	185.62	9.40	0.95
10	103.8	11.27	7.58	205.14	9.39	0.00
11	96.7	14.88	2.92	124.93	12.30	7.55
12	99.3	11.27	3.69	122.72	12.79	7.65
13	99.3	3.16	7.87	165.55	8.98	0.83
14	99.1	6.36	5.04	122.98	13.33	7.44
15	99.5	6.71	7.97	202.69	9.70	1.08
16	71.5	1.27	1.32	121.92	17.34	15.80
17	84.1	33.87	0.07	171.10	14.08	9.26
18	80.1	90.05	0.74	*	12.51	0.51
19	83.2	100.00	0.13	*	**	0.00
20	98.9	14.73	3.41	129.52	13.31	7.84
21	97.6	18.64	6.57	223.36	9.35	1.06
22	90.4	24.40	1.65	140.51	13.96	8.48
23	88.1	87.20	1.02	183.20***	9.41	0.23

* not determined
** no residue
*** neutralization equivalent of the extraction residue

are shown in Figure 5. Comparison of these spectra with
that of phthalamic acid (B in Figure 2) show that a very
significant change has occurred during the reaction.
 After reation at 150°C (spectrum A), the N-H stretching
doublet has been replaced by a strong absorption at 3125
cm^{-1} and a weak absorption at 3480 cm^{-1}. The carboxyl
stretching region has also changed. The amide and
carboxyl absorbances at 1695 cm^{-1} and 1645 cm^{-1} have
been replaced by two peaks at 1675 cm^{-1} and 1550 cm^{-1}.
The latter peak is quite intense. Additionally, two
broad, moderately intense peaks have appeared at 2450 cm^{-1}
and 1950 cm^{-1}. The absorbances are consistent with
those of an ammonium salt containing some dicarboxylic
acid as noted before in the discussion of the reaction
of succinamic acid. In this case, however, the spectrum
indicates the hydrolysis reaction is virtually
complete.
 With increasing temperature, the spectra change,
particularly in the carbonyl absorption region (spectra
B, C). The peak which appears at 1675 cm^{-1} in
spectrum A has shifted to 1705 cm^{-1} in spectrum B and
has increased in intensity relative to the maximum at
1550 cm^{-1}. In addition a small shoulder has appeared
at 1780 cm^{-1} in spectrum B. In spectrum C, the high
energy maximum now is the most intense and is located
at 1745 cm^{-1}. The shoulder at 1780 cm^{-1} in B is now a
sharp peak at 1780 cm^{-1}. In addition the broad peaks
at 2450 cm^{-1} and 1950 cm^{-1} noted in A have
substantially disappeared in C. The N-H stretching
region is virtually the same in each spectrum but, as
noted before, differs radically from that of phthalamic
acid. These spectra are consistent with the conclusion
that phthalamic acid is almost completely hydrolyzed in
the pad-bake reaction or during the preparation of the
pad solution. The material deposited is essentially
the monoammonium salt of phthalic acid. This salt is
increasingly converted to the inactive product,
phthalimide as the temperature is increased.
 These conclusions are supported by the analytical
data in Table II. The % amide ammonia is only 0.31% in
the products of reaction 4 and zero in reactions 5 and
6. The ionic ammonia values are relatively high but
decrease from 8.36% at 150°C (reaction 4) to 5.16% at
200°C (reaction 6). The neutralization equivalent and
% chloroform soluble material increase from 196.67
grams/equivalent and 6.95% at 150°C (reaction 4) to
331.85 grams/equivalent and 45.70% at 200°C (reaction
6). These observations can be explained by the
progressive conversion of the monoammonium salt into
phthalimide. By using the methods discussed previously

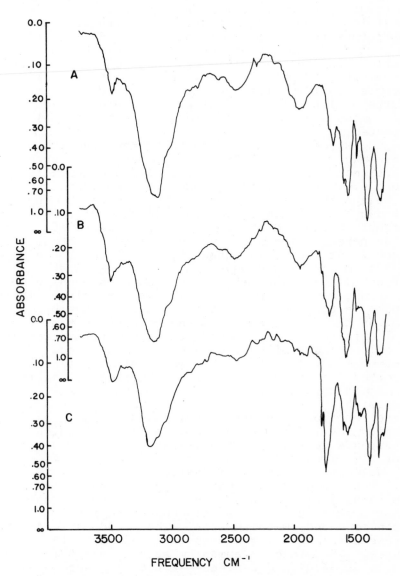

Figure 5.   Infrared spectra of the products of the reaction of phthalamic acid for 10 min at 150°C (A), 175°C (B), and 200°C (C)

the yields of the various products can be calculated.
The results of the calculations are shown in Table III.
The yield of ammonium salt decreases with increasing
temperature from an initial value of 87.5% at 150°C.
The amic acid was almost entirely hydrolyzed in these
reactions and the hydrolyzate, monoammonium phthalate,
is converted to phthalimide. The yield of the imide
increases rapidly with temperature from 8.4% after 10
minutes at 150°C to 48.7% after 10 minutes at 200°C.
Small amounts of the amic acid and diacid were also
indicated.

The spectra of the products of the reaction of
succinamic acid at 175°C for 5, 10, and 15 minutes
(reactions 7, 2, and 8, respectively) are shown in
Figure 6. With increasing bake time the changes noted
in the spectra are very nearly the same as the changes
caused by increasing temperature (Figure 3). The amide
carbonyl stretching absorption which appears at1650 cm^{-1}
in the parent compound, succinamic acid (spectrum A,
Figure 2) decreases in intensity with time as the peak
at 1575 cm^{-1} broadens and a shoulder appears at
1760 cm^{-1}. Again the nitrogen-hydrogen stretching
vibrations have reversed in relative intensity. These
changes in the spectra of the products of the bake
reaction are consistent with the conclusion that the
succinamic acid is deposited during volatilization of
the solvent as a mixture of succinamic acid and
monoammonium succinate. This material is progressively
converted into succinimide as the heating is continued.

Again the analytical data support   this
conclusion. Five minutes into the reaction (reaction
7) the reaction products contain 3.63% ammonia as
ammonium ion. The % amide ammonia is 9.61% which
corresponds to a yield of 65.9% succinamic acid. Thus,
after five minutes, the amount of amic acid has
decreased 34.1%. The yield values are given in Table
III. The yield of succinamic acid decreases with time
from 65.9% after 5 minutes to 39.8% after 15 minutes as
the amic acid is progressively converted into
succinimide and ammonium salt. The yield of imide
increases from 0.6% at 5 minutes to 23.1% at 15
minutes. The yield of ammonium salt is maximum at 10
minutes at 34.0% and decreases after 15 minutes to
23.1% indicating the possibility of its conversion to
diacid or imide. The yield of diacid increased from
4% after 5 minutes to 13% at 15 minutes.

The spectra of the products of the reaction of
phthalamic acid at 175°C for 5, 10, and 15 minutes
(reactions 9, 5, and 10 respectively) are shown in
Figure 7. The spectra again differ greatly from the

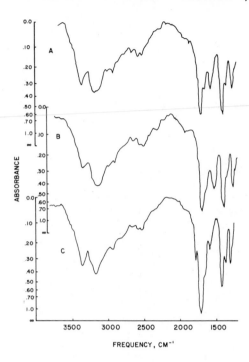

Figure 6. *Infrared spectra of the products of the reaction of succinamic acid at 175°C for 5 min (A), 10 min (B), and 15 min (C)*

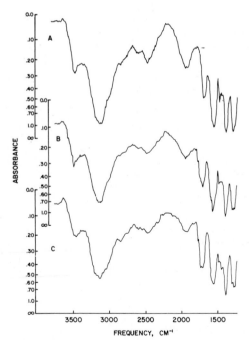

Figure 7. *Infrared spectra of the products of the reaction of phthalamic acid at 175°C for 5 min (A), 10 min (B), and 15 min (C)*

Table III

Product Yield Data, Time, Temp., Conc. and Catalyst

% Yield

Reaction Number	Unreacted Amic Acid	Mono-ammonium salt	Imide	Diacid
1	64.6	28.6	0.5	6.
2	53.2	34.0	5.7	6.
3	41.5	7.7	41.1	11
4	3.3	87.5	8.4	0
5	0	81.5	14.9	1
6	0	47.6	48.7	1
7	65.9	24.7	0.6	4
8	39.8	23.1	23.1	13
9	10.2	88.6	2.5	0
10	1.4	82.1	17.2	0
11	50.2	19.3	17.0	8
12	52.1	25.0	13.1	9
13	7.8	74.6	3.6	10
14	51.3	34.6	7.6	7
15	10.9	80.3	6.7	1

spectrum of phthalamic acid (Figure 2, spectrum B) and
indicate by the absence of any significant absorption
at 1645 cm^{-1} due to the amide carbonyl stretching
absorption and by the presence of the strong absorption
at 1550 cm^{-1} and 3200 cm^{-1} that the phthalamic acid is
extensively hydrolyzed. Dehydrocyclization to yield
imide also occurs as indicated by the appearance of the
shoulder at 1780 cm^{-1} in spectra B (reaction 5) and C
(reaction 10). These observations indicate that the
phthalamic acid is almost completely hydrolyzed at the
outset of the bake step and is deposited during the
volatilization of the solvent as the partial ammonium
salt. The analytical data in Table II support this
conclusion with the % ionic ammonia quite high at
8.25% after 5 minutes (reaction 9) and decreasing to
7.58% after 15 minutes (reaction 10). This decrease is
accompanied by an increase in the % chloroform
extractables from 2.05% to 11.27% and the
neutralization equivalent from 185.62 gram/equivalent
to 205.14 gram/equivalent in the same reactions. These
data were used to calculate the % yield values shown in
Table III. Here the yield of ammonium salt is quite
high at 88.6% after five minutes and decreases to a
nearly constant value of 81.5% after 10 minutes. The
yield of phthalimide increases continuously throughout
the reaction from a value of 2.5% after 5 minutes to
17.2% after 15 minutes. The residual phthalamic acid
yield which is 10.2% after 5 minutes has decreased to
1.4% after 15 minutes. Phthalic acid was detected at
10 minutes in a 1% yield. These data indicate that the
ammonium salt is formed in high yield during the
preparation of the amic acid solution and during the
volatilization of the solvent by the hydrolysis of a
substantial amount (88.6% at 5 minutes) of the initial
phthalamic acid. The residual phthalamic acid disappears gradually as does the ammonium salt. This
disappearance is accompanied by the appearance of
phthalimide which could result from the decomposition
of the phthalamic acid and the partial ammonium salt.
     It is of interest to determine the effect of the
pad bath concentration on the residue composition. The
spectra of the products of the reaction of succinamic
acid at 175°C for 10 minutes at initial concentrations
of 4.00, 2.50, and 1.50 molar (reactions 12, 2, and 11,
respectively) are shown in Figure 8. In comparison
with the spectrum of succinamic acid (spectrum A in
Figure 2) all the spectra show a decrease in the amide
carbonyl stretching absorption at 1650 cm^{-1} accompanied
by broadening and increase in absorbance at 1575 cm^{-1},
the carboxylate carbonyl stretching absorption, a

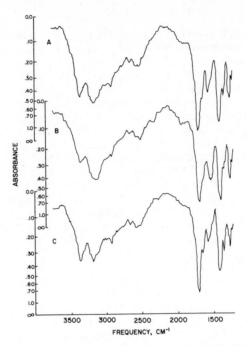

Figure 8. Infrared spectra of the products of the reaction of succinamic acid at 175°C for 10 min at an initial solution concentration of 4.00M (A), 2.5M (B), and 1.50M (C)

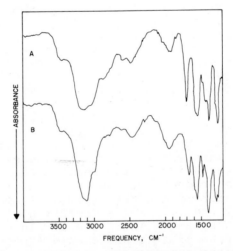

Figure 9. Infrared spectra of the products of the reaction of phthalamic acid at 175°C for 10 min at an initial solution concentration of 1.50M (A) and 2.50M (B)

change in the relative intensity of the doublet in the
nitrogen-hydrogen stretching region (3000 cm^{-1} to
3500 cm^{-1}), and the appearance of a shoulder at
1760 cm^{-1}. These changes indicate that the amic acid
has undergone both hydrolysis and dehydrocyclization.
From the spectra it can be seen that the extent of
reaction is greatest for reaction 2 (spectrum B) since
the spectrum of the residue of this reaction shows the
greatest change. The analytical data in Table II show
quantitatively the changes which occur in the succinamic acid during the pad-bake reaction. The amount of
hydrolysis which has occurred is the greatest at the
intermediate concentration as indicated by the % ionic
ammonia values. In reactions 11 and 12, the values
for the % ionic ammonia are 2.92% and 3.69%
respectively while in reaction 2, the value is 5.27%.
The % chloroform extractables is lower at the
intermediate concentrations where the value is 5.84%
for an initial concentration of 2.50 molar (reaction 2)
and 14.88% and 11.27% at initial concentrations of 1.50
(reaction 11) and 4.00 molar (reaction 12),
respectively. These results indicate that the extent
of dehydrocyclization reaction is less than that at the
other concentrations. The neutralization equivalents
of the residues are nearly the same as are the % amide
ammonia values for all the initial concentrations. The
yield data calculated as before are shown in Table III.
The data show little effect of the initial concentra-
tration on the yield of succinamic acid in the range of
1.50 to 4.00 molar after 10 minutes bake time at 175°C.
 The yield of imide generally decreases with the
solution concentration while the yield of ammonium salt
increases. The yield of succinic acid varies between
6% and 9%.
      The spectra of the products of the reaction of
phthalamic acid at 175°C for 10 minutes at an initial
concentration of 1.50 molar (reaction 13) and 2.50
molar (reaction 5) are shown in Figure 9. (Solutions
of phthalamic acid of a concentration of 4.00 molar
could not be prepared). The spectra show that the
phthalamic acid has been extensively hydrolyzed and
that dehydrocyclization has taken place to a slight
extent as might be expected from the previous results.
In Table II the data show that the % ionic ammonia
values for reactions 5 and 13 are 8.08% and 7.87%
respectively indicating the extent of hydrolysis is
about the same in each case. At the lower initial
(1.50 molar) concentration the % chloroform
extractables is 3.16% compared with 12.41% at an
initial concentration of 2.50 molar. The neutrali-

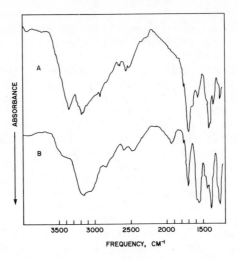

Figure 10. Infrared spectra of the products of the reaction of succinamic and phthalamic acids at 175°C for 10 min at an initial concentration of 2.50M and 0.171M ammonium sulfamate as catalyst

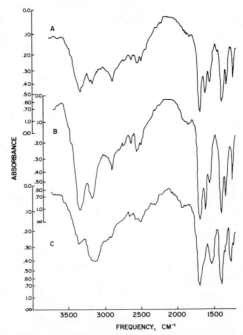

Figure 11. Infrared spectra of the products of the reaction of succinamic acid at 175°C for 10 min at an initial concentration of 2.50M in N,N-dimethylformamide (A), formamide (B), and water (C)

zation equivalent of the residue in reaction 5
is 204.41 grams/equivalent while that of the residue in
reaction 13 is 165.55 grams/equivalent. The % amide
ammonia values are 0.00 and 0.83% for reactions 5 and
13 respectively. These data were used to calculate the
yield values given in Table III which show by the yield
of ammonium phthalate that the amount of hydrolysis is
about the same in each reaction. The yield of diacid
decreases with initial concentration while the yield of
imide increases. The yield of residual phthalamic acid
remains about the same.

The spectra of the products of the reactions of
succinamic acid (spectrum A, reaction 14) and
phthalamic acid (spectrum B, reaction 15) at 175°C for
10 minutes at an initial concentration of 2.50 molar in
amic acid and 0.171 molar in ammonium sulfamate
catalyst are shown in Figure 10. The spectra differ in
the same manner from the spectra of the parent
compounds, succinamic acid (Figure 2, spectrum A) and
phthalamic acid (Figure 2, spectrum B) as do the
products of reaction 2 (Figure 3, Spectrum B) and
reaction 5 (Figure 6, spectrum B) which were performed
under identical conditions as reactions 14 and 15
except for the presence of the catalyst. The
analytical data in Tale II reflect the similarities.

The % ionic ammonia in reaction 2 is 5.27% compared
with 5.04% in reaction 14. In reaction 5 the % ionic
ammonia is 8.08% whereas in reaction 15 the value is
7.97%. In reaction 2 the values for % chloroform
extractables, neutralization equivalent, and % amide
ammonia were 5.84%, 123.13 grams/equivalent, and 7.35%,
respectively, while in reaction 14 with catalyst
present the values were 6.36%, 122.98 grams/equivalent,
and 7.44%, respectively. In reaction 5 the values for
% chloroform extractables, neutralization equivalent,
and % amide ammonia were 12.41%, 204.41
grams/equivalent, and 0.00% respectively while in
reaction 15 the values were 6.71%, 202.69
grams/equivalent, and 1.08%. The analytical data are
translated into product yields in Table III. Clearly
the presence of ammonium sulfamate catalyst has no
great effect on the yields of the products in the
residue.

The spectra of the products of reaction of
succinamic acid in various solvents are shown in
Figure 11. The solutions were 2.50 molar in
N,N-dimethylformamide (reaction 17, spectrum A),
formamide (reaction 16, spectrum B), and water
(reaction 2, spectrum C). Comparison of the spectra in
Figure 11 with that of the pure succinamic acid in

Figure 2 (spectrum A) reveals a greater difference in
reaction 2 with water as solvent than in reactions 16
and 17 with nonaqueous solvents. The spectrum of the
products of reaction 16 (B) is very similar to that of
the pure succinamic acid with only a slight decrease in
the amide carbonyl absorption. In spectrum A this
difference is more pronounced. Apparently, the changes
taking place in DMF and formamide are quite different
from those occurring in water. The analytical data in
Table II reflect these differences. The % ionic
ammonia values for the products from formamide
(reaction 16) and DMF (reaction 17) are 1.32% and
0.07%, respectively, whereas the value is 5.27% for
water (reaction 2). This indicates that a much
greater amount of hydrolysis takes place in water than
in the other solvents, as expected. The % chloroform
extractables is 5.84% for reaction 2 (water), 1.27% for
reaction 16 (formamide), and 33.87% for reaction 17
(DMF). This indicates that imide is formed more easily
in the aprotic solvent, DMF, than in the protic
solvents, water and formamide. The neutralization
equivalent and the % amide ammonia values of the
products of the reactions are 123.13 grams/equivalent,
7.35%; 121.92 grams equivalent, 15.80% and 171.10
grams/equivalent, 9.26% for reactions 2, 16, and 17,
respectively. The data were used to calculate the
product yields in Table IV. The amount of the original
succinamic acid still present after the pad-bake
procedure is much greater for formamide (76.6%) than
for DMF (59.3%) and water (53.2%). The hydrolysis
reaction takes place to a much greater extent in water
than in DMF or formamide as indicated by the yields of
monoammonium succinate. These yield values are 34.0%,
7.4% and 0.4% for water, formamide, and DMF,
respectively. The dehydrocyclization reaction to yield
succinimide occurred to a much greater extent in DMF
(37.3% yield) than in water (5.7%) or in formamide
(1.0%). Succinic acid was detected in low yield in
reactions 2 and 17. The yield of succinic acid was 6%,
0%, and 2% for reactions 2, 16, and 17, respectively.
      The spectra of the products of the reaction of
phthalamic acid in DMF (reaction 19), formamide
(reaction 18), and water (reaction 5) are shown in
Figure 12. The reaction conditions employed were 10
minutes bake time at 175°C and a solution concentration
of 2.50 molar. The spectra of the reaction products
differ greatly from that of the pure phthalamic acid
(Figure 2, spectrum B). The spectra indicate that with
formamide (reaction 18, spectrum B) and DMF (reaction
19, spectrum A) the residue spectra are quite similar

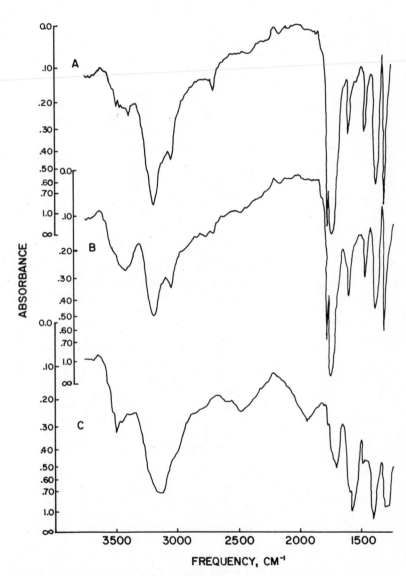

Figure 12.   Infrared spectra of the products of the reaction of phthalamic acid at 175°C for 10 min at an initial concentration of 2.50M in N,N-dimethylformamide (A), formamide (B), and water (C)

and indicate that the major product is phthalimide. In
water (spectrum C) the major product is the ammonium
salt. The data in Table II show that the % ionic
ammonia vlues for the products of reactions 18 and 19
are only 0.74% and 0.13%, respectively, while that of
reaction 5 is 8.08%. The % chloroform extractables is
quite high at 90.05% in reaction 18 and 100.00% in
reaction 19. The values would be expected to be large
due to the large amount of imide in the residue. The %
amide values are 0.51% for reaction 18 and zero for
reaction 19. The product yield values are given in
Table IV. The data show that in all three solvents the
initial phthalamic acid is almost completely converted
to one major product. In water (reaction 5) this
product is the monoammonium salt of phthalic acid
(81.5% yield) while in formamide (reaction 18) and DMF
(reaction 19) this product is phthalimide (74.7% and
92.5% yield, respectively).

Allen has shown that the manner in which the
solvent is removed in the pad-bake reaction of aqueous
succinamic acid solutions with viscose rayon greatly
affects the extent of the esterification reaction (4).
Fabric which was padded, dried at room temperature and
baked was esterified to a much lesser extent than
fabric which was dried and baked at an elevated
temperature in the same step. Allen also found that
little reaction occurred between molten succinamic acid
and viscose rayon fabric, in essence, a reaction in
which the amic acid is the solvent. To determine the
effect of the solvent removal conditions on the
composition of the residue deposited on the fabric,
amic acid was reacted under three sets of conditions:
normal pad-bake, pad, predry at room temperature for 24
hours under a stream of air and bake, and fusion
without added solvent.

The spectra of the products of the reaction of
succinamic acid under normal (reaction 2) predry
(reaction 20) and fusion conditions (reaction 22) are
shown in Figure 13. In each reaction, the spectrum of
the product differs significantly from that of the pure
succinamic acid (Figure 2, spectrum A). In all the
spectra the amide carbonyl stretching absorption has
decreased significantly, indicating partial conversion
of the original succinamic acid into succinimide and
monoammonium succinate. The analytical data in Table
II reveal that the % ionic ammonia is 5.27%, 3.41%, and
1.65% in reactions 2, 20, and 22, respectively. The
amount of hydrolysis which has taken place is greatest
in the pad-bake procedure and least in the fusion
procedure. The % chloroform extractable values for

Table IV

Product Yield Data, Solvents

Reaction Number	Unreacted Amic Acid	Mono-ammonium salt	Imide	Diacid
16	76.7	7.4	1.0	0
17	59.3	0.4	37.3	2
18	3.9	5.7	74.7	0
19	0	1.1	92.5	0
20	53.1	23.1	17.3	9
21	10.0	56.3	28.2	0
22	52.7	13.6	30.8	2
23	2.0	8.7	85.8	0

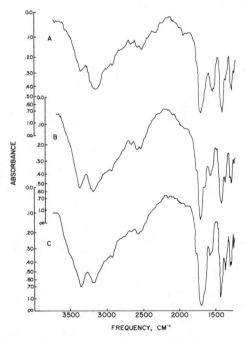

Figure 13. Infrared spectra of the products of the reaction of succinamic acid at 175°C for 10 min under normal (A), predry (B), and fusion (C) conditions

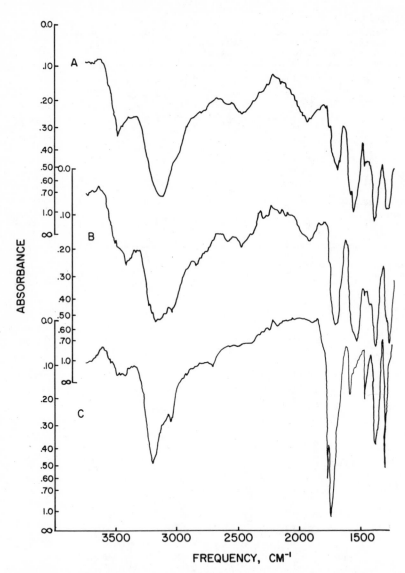

Figure 14. Infrared spectra of the products of the reaction of phthalamic acid at 175°C for 10 min under normal (A), predry (B), and fusion (C) conditions

reactions 2, 20, and 22 are 5.84%, 14.73% and 24.40%,
respectively, indicating that imide formation is the
most important in the fusion reaction (22). This is
also reflected in the values of the neutralization
equivalents, 123.1, 129.5, and 140.5 grams/equivalent
for reactions 2, 20, and 22, respectively. The
analytical data were used to calculate the % yield
values in Table IV. The % yield of succinamic acid is
about the same in each experiment. The yield of
monoammonium salt is greatest in the normal pad-bake
procedure (34.0%) at least in the fusion procedure
(13.6%). This relationship is reversed in the % yield
of succinimide with the normal procedure yield 5.7% and
the fusion procedure yield 30.8%. With the predry
conditions the yields of monoammonium salt and
succinimide are intermediate in value at 23.1% and
17.3%, respectively. These results indicate that
hydrolysis of the amic acid is most complete in the
reaction with water as solvent and that the higher the
temperature of removal of the water the greater the
hydrolysis. Dehydrocyclization to form imide is most
favored in the absence of water as the results of the
fusion reaction indicate. The yield of succinic acid
was 6%, 9%, and 2% in the normal, predry, and fusion
reactions respectively.

The trends noted in the succinamic acid reaction
above were even more apparent in the reactions of
phthalamic acid under normal, predry, and fusion
conditions. The spectra of the products of the
reactions of phthalamic acid under normal (reaction
5), predry (reaction 21), and fusion conditions
(reaction 23) are given in Figure 14 (spectra A, B, and
C, respectively). Spectra A and B show that hydrolysis
is the predominant reaction by the absence of the amide
carbonyl stretch at 1645 cm^{-1} and the presence of the
strong carboxylate carbonyl stretch at 1550 cm^{-1} .
Spectrum C indicates that the predominant product in
the fusion reaction is phthalimide. These conlusions
are reflected in the analytical data in Table II. In
reactions 5 and 21 the % ionic ammonia in the product
is 8.08% and 6.57%, respectively, while in reaction 23
the value is 1.02%. The % chloroform extractable
values are 12.41% and 18.64% in reactions 5 and 21,
respectively, and 87.20% in reaction 22. The
neutralization equivalent values for reactions 5 and 21
were 204.4 and 223.4 grams/equivalent, respectively.
The neutralization equivalent of the products of
reaction 23 was not directly determined. The residue
of the chloroform extraction was found to have a
neutralization equivalent of 183.2 grams/equivalent.

Table V

Mass Balance and Volatile Product Yield Data

Reaction Number	% Yield of Acid Derivatives	% Yield of Nitrogen Cont. Cmpds.	% Yield of Volatile Ammonia	% Yield of Sublimate
1	99.7	93.7	0.0	
2	98.9	92.9	0.8	
3	97.1	87.1	4.1	
4	99.2	99.2	0.2	
5	98.0	97.0	0.2	
6	97.3	96.3	0.3	
7	95.2	91.0	0.0	0.2
8	99.0	86.0	0.6	0.8*
9	100.5	100.5	0.1	0.5**
10	100.8	97.0	0.0	0.8**
11	94.5	86.5	4.0	
12	99.4	90.4	2.3	
13	96.0	86.0	0.8	
14	100.4	94.4	1.7	
15	100.0	99.0	0.0	
16	86.1	85.1	0.6	
17	99.0	97.0	1.7	
18	84.4	84.4	3.9	
19	93.6	93.6	5.9	
20	98.0	89.0	2.2	
21	92.7	93.3	0.0	
22	98.6	94.6	0.6	
23	96.5	96.5	0.1	

* As succinamic acid
** As phthalimide

The % amide ammonia was greatest in reaction 21 at
1.06% and least in reaction 5 at 0.10%. The product
yield values are given in Table IV. The phthalamic
acid is almost totally converted into products under
all three sets of conditions. The reactions under
normal and predry conditions give the most hydrolysis
as indicated by the % yield of monoammonium salt, 81.5%
and 56.3%, respectively. In the fusion reaction, 23,
the imide yield is 85.8% while in the normal (reaction
5) and predry reactions the yields are 14.9% and 28.2%,
respectively. Little phthalic acid was detected.

Volatile Products

During the pad-bake procedure volatile products
other than the solvent are evolved. This becomes
apparent when the material balance for the reactions is
computed. The material balance on the acid derivatives
and nitrogen containing products is given in Table V
along with the experimentally determined values of the
yields of volatile products. The material balance
values vary from 100.5% in reaction 9 to 84.4% in
reaction 18. In the reactions in formamide (16 and 18)
the solvent was not completely removed in the bake step
due to the low volatility of the solvent. The solvent
was removed by evaporation at room temperature under
high vacuum for several hours. It is possible that
under these conditions a portion of the products
sublimed from the mixture and was lost. In both
reactions 16 and 18 the material balance values are
comparatively low at 86.1% and 84.4%, respectively, for
the acid derivatives, and 85.1% and 84.4% for the
nitrogen containing products. The mass balance values
decrease with both the duration and temperature of
reaction. The % yield of succinic acid derivatives and
nitrogen containing compounds is 99.7% and 93.7%,
respectively in reaction 1 (150°C bake temperature)
while in reaction 3 (200°C bake temperature) the values
are 97.1% and 87.1%, respectively. The yield of
phthalamic acid derivatives and nitrogen containing
compounds was 99.2%, and 99.2%, respectively, in
reaction 4 (150°C bake temperature), and 97.3% and
96.3%, respectively, in reaction 6 (200° C bake
temperaure).

The experimentally determined yields of sublimate
and volatile ammonia are also given in Table V. The
sublimate was obtained by evaporation of the trap of
chloroform in reactions 8, 9, and 10. On evaporation
of the chloroform a small amount of residue was
obtained. The infrared spectra of the residues are

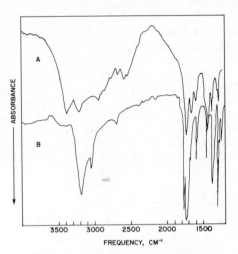

Figure 15.   Infrared spectra of the volitile products in reactions
8 (A) and 9 (B)

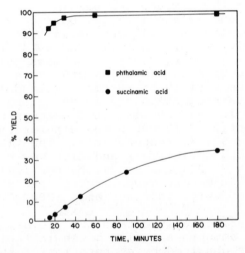

Figure 16.   Variation of the percent yield of ammonium ion in
the padbath with time for 2.50M solutions of succinamic and
phthalamic acid at 65° ± 1°C

given in Figure 15.   Spectrum A, that of the sublimate
from reaction 8, when compared with that of the pure
compound in Figure 2 (spectrum A) appears to be the
spectrum of a partially hydrolyzed or neutralized
sample of succinamic acid.   Spectrum B, obtained from
the product of reaction 9 is identical to the spectrum
of phthalimide.   The same is true for the spectrum of
the volatile product of reaction 10 (not shown).   Thus,
the material which sublimes from the reaction is
succinamic acid in the reactions of succinamic acid and
phthalimide in the reaction of phthalamic acid.

The % yield of volatile ammonia in the reactions
of succinamic acid is probably too low since the
succinamic acid itself is also passed into the trap of
standard acid as it sublimes.   In the reactions of
phthalamic acid in water the yield of volatile ammonia
is generally lower than the yield of volatile ammonia
in the corresponding succinamic acid reaction.   The
exception to this is in the reactions in formamide and
DMF.   Succinamic acid in formamide (reaction 16) and
DMF (reaction 17) yields 0.6% and 1.7%, respectively,
of volatile ammonia, and phthalamic acid in formamide
(reaction 18) and DMF (reaction 19) yields 3.9% and
5.9% ionic ammonia, respectively.

Pad Bath Stability

It is of interest to determine the changes which
occur in the pad-bath before the bake step.   The main
change expected in the pad bath is hydrolysis of the
amic acid.   This reaction was followed by determination
of the ionic ammonia content of a 2.50 molar aqueous
pad-bath held at 65°C and by following the changes in
the carbonyl stretching region of the infrared spectrum
of the material in the bath.   The % yield of the
ammonium ion present in the pad bath is shown as a
function of time in Figure 16.   The time of
introduction of the amic acid into the water was the
beginning of the reaction.   Complete dissolution of the
amic acid occurred within five minutes in the
procedure.   Phthalamic acid is almost entirely
hydrolyzed during the dissolution period as hydrolysis
rapidly occurs.   Pad-bake reactions with aqueous
solutions of phthalamic acid certainly involve the
monoammonium salt.   Succinamic acid is more resistant
to hydrolysis with a yield of monoammonium succinate of
only 34.0% after 3 hours.   These changes are also
reflected in the changes which occur in the carbonyl
stretching region of the spectra of the material in the
bath which are shown in Figure 17.   The spectra show

that the amide carbonyl stretching absorption at 1650
cm^{-1} gradually disappears as the carboxylate carbonyl
stretching absorption gradually appears at 1550 cm^{-1}
The changes in the spectra of the phthalamic acid
residues are more dramatic than those in the
succinamic acid experiment. The amide carbonyl
absorption at 1645 cm^{-1} is completely absent at 15
minutes while the carboxylate carbonyl stretching
absorption is fully developed in 15 minutes. There is
little further change in the spectra with time.

Conclusions

During the pad-bake process the amic acid is
dissolved in the solvent at an elevated temperature.
The resulting solution is applied to the substrate, the
solvent is rapidly volatilized, and a residue is
deposited on the substrate. The residue is then
further heated in the completion of the bake step. The
process can for the sake of discussion be divided into
three steps: solution preparation and padding,
volatilization of solvent, and bake of the reactants.
During the preparation of the solution the
predominant chemical change which occurs in the amic
acid is hydrolysis. This was graphically illustrated
in the experiments to determine the pad-bath stability.
Phthalamic acid is almost entirely hydrolyzed during
the preparation of the solution while succinamic acid
is hydrolyzed at a somewhat slower rate. This result
is not surprising in light of the studies of the
hydrolysis of , α β-amic acids which have been
reported (10,11,12,13,14,15,16,17,18). α, β -amic acids
which are derived from dicarboxylic acids forming
internal pentacyclic anhydrides are hydrolyzed at rates
which are uncharacteristically high for amides. The
mechanism first proposed to explain this unusual
behavior involves the formation of the internal
anhydride in a rate determining step followed by rapid
hydrolysis (10,11). This mechanism was inferred from
the results of an isotopic labeling experiment (11).
The facile hydrolysis of many α, β -amic acids has
since been reported (12,13,14,15,16,17,18), and the
anhydride was directly detected as the reactive
intermediate in the reaction in one recent study (18).
The rate constants for the pseudo first order
hydrolysis of succinamic and phthalamic acids at 65 °C
are 3.3 x 10^{-5} sec^{-1} and 1.3 x 10^{-3} sec^{-1},
respectively (12). By use of these first order rate
constants, the concentration of amic acid in the pad
bath at various times can be estimated. These values

Table VI

Comparison of theoretical and experimental
values of % yield of ammonium salt during
hydroylsis of the amic acids in the pad bath

Age of the Pad Bath (minutes at 65°C)	% Yield of Ammonium Salt in Pad Bath			
	Succinamic Acid		Phthalamic Acid	
	Found	Calculated[a]	Found	Calculated[a]
15	1.5	3.0	92.2	87.8
20	3.0	3.9	95.0	94.0
30	6.5	5.8	97.5	98.5
45	12.6	8.6		
60		11.3	98.5	99.9
120	23.2	21.3		
180	33.8	30.2	99.0	100.0
360		51.3		

[a]From the first order rate constant in reference (12).

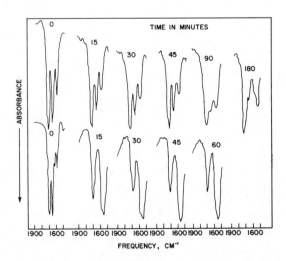

Figure 17. Variation in the carbonyl stretching absorption region of the infrared spectra of the pad-bath residues with time for 2.50M solutions of succinamic and phthalamic acid at 65° ± 1°C

are shown in Tale VI along with experimental values
from the determination of the pad-bath stability. The
data from this study parallel closely the calculated
data. Also, the rate constant for the pseudo first
order hydrolysis of succinamic acid was calculated to
be 2.6 x 10^{-5} sec^{-1} which is in good agreement with
the value of Kezdy and Bruylants (12) given above. The
value of the rate constant for the phthalamic acid
could not be estimated from the experimental data. The
close parllel between the data obtained in this study
to that predicted by the literature indicate a
similarity in the paths by which the amic acid is
hydrolyzed.

     Notably, the two amic acids differ greatly in
their stabilities as this study indicates. The
difference in the reacivity of the two amic acids can
be readily explained in terms of their structures. The
conformationally mobile succinamic acid can exist as
extended chain conformers which do not favor the
carboxyl-amide interaction demanded by the Bender
mechanism for facile hydrolysis (11). The phthalamic
acid molecule is stucturally rigid and the carboxyl and
amide groups are constrained to interact. The relative
rates of hydrolysis of the two compounds at 65°C is
given by the ratio of the above rate constants (11),
thus phthalamic acid is hydrolyzed 39 times faster than
succinamic acid. Other structurally rigid α , β -amic
acid such as maleamic acid would be expected to behave
in the same manner as phthalamic acid and be
substantially hydrolyzed during the preparation of the
pad-bath at 65°C. The rate constant for the hydrolysis
of maleamic acid at 65°C is 2.4 x 10^{-3} sec^{-1} or roughly
twice that of phthalamic acid (11). Calculations using
this constant predict that maleamic acid would be 88%
hydrolyzed in 15 minutes in an aqueous pad-bath at 65°C

     In the second step in the pad-bake procedure the
solvent is volatilized. In the experiments with water,
the volatilization of the bulk of the solvent occurred
in approximately two minutes for the 2.50 molar
solutions. Formamide and DMF, less volatile solvents
than water were present throughout the reaction. Bake
temperature made little apparent difference in the time
required for the removal of the bulk of the water.
The results of the experiments with a bake time of 5
minutes at 175°C , reaction 7 for succinamic acid and
reaction 9 for phthalamic acid, give some insight into
the changes which occur in the amic acid during
volatilization of the solvent.

     In reaction 7 only 65.9% of the initially present
amide was accounted for by the amide ammonia analysis.

Since succinamic acid is relatively unchanged in the
bath after 10 minutes, the maximum bath age at the time
of withdrawal of the aliquot of solution for reaction,
the extent of hydrolysis of succinamic acid would be
relatively small as predicted by the data in Table IV
and Figure 16. Consequently, a substantial amount of
hydrolysis must occur during the volatilization step.
By using the activation parameters reported in reference 12 the rate constant for the hydrolysis of
succinamic acid in water at 100°C can be shown to be
6.9×10^{-4} sec^{-1}. If hydrolysis proceeds for two minutes
at this rate the yield of ionic ammonia expected would
be at most about 8-10%, depending on the amount of
hydrolysis occurring in the solution preparation step.
The yield of ionic ammonia in reaction 7 was found to
be 24.7%, more than twice the maximum amount predicted
by the first order hydrolysis constant. This
discrepancy could arise for a number of reasons. The
temperature in the reaction mass probably deviates
considerably from that of boiling water, particularly
during the latter stages of the volatilization when the
solution is concentrated. Also, as water is removed
the reaction medium undergoes a drastic change from
aqueous solution to a solution of the products of the
reactions in liquid succinamic acid. Under these
conditions the pseudo first order rate constants
probably are not valid. Infrared studies of solutions
of N-substituted succinamic acids in non-polar or
aprotic solvents and in the solid state indicate the
formation of hydrogen bonded dimers (33,32). In one
study the dimeric form of the amic acid was postulated
to be more reactive in solvolysis reactions (22). The
reaction rate increased with an increase in the
concentration of succinamic acid or on the addition of
acetic acid. Sauers, et . al. (25), also noted an
effect of added acid in the attempted
dehydrocyclizations of N-substituted phthalamic acids
in acetic anhydride. In their study the competition
between dehydrocyclization to form imides or isoimides
and the internal transacylation to form phthalic
anhydride favored anhydride formation in the presence
of added acid or large initial concentrations of amic
acid. The mechanism proposed to explain these
observations involved intermediate formation of a mixed
phthalic acid-acetic acid anhydride or direct formation
of phthalic anhydride by a pathway similar to the
Bender mechanism (11). Thus, the observations noted in
the literature lend support to the conclusion that the
rate of hydrolysis of succinamic acid could be enhanced
during volatilization of the water due to the change

in reaction medium.

With phthalamic acid, little change occurs during
the volatilization of the solvent since the compound is
substantially hydrolyzed during the preparation of the
pad-bath. In reaction 9, the yield of ionic ammonia
was 88.6%, while in the pad bath after 15 minutes the
yield of ionic ammonia was 92.2%. The residue
deposited by volatilization of the solvent is solid and
consists principally of monoammonium phthalate.
Reactions of aqueous solutions of phthalamic acid with
viscose rayon generally lead to lower D.S. values than
does the same reaction with succinamic acid (32). Our
results indicate that such reactions proceed through
the partial ammonium salt which is deposited as a solid
and consequently has low mobility in any further
reaction. This probably accounts in part for the low
reactivity of phthalamic acid in its reactions with
cellulosic substrates.

In the bake step the residue deposited by
volatilization of the solvent undergoes further
chemical change. The predominant reaction during this
phase of the reaction with both succinamic and
phthalamic acid is dehydrocyclization to yield imide.
Since neither succinimide nor phthalimide (3) undergoes
any significant reaction with cellulosic substrates in
pad-bake reactions, the dehydrocyclization reaction
represents a serious competition in the formation of
cellulose esters by reaction with α, β -amic acids.

After 5 minutes at 175°C (reaction 7) only 65.9%
of the original succinamic acid remained in the residue
as measured by the amide ammonia determination. The
majority of the succinamic acid which has reacted has
undergone hydrolysis as indicated by the high yield of
ionic ammonia (24.7%) in the residue. At ten minutes
bake time (reaction 2) the succinamic acid content has
further decreased to 53.2% while the yield of both
imide and ionic ammonia have increased to 5.7% and
34.0%, respectively. This indicates that succinamic
acid is further hydrolyzed in the melt deposited even
after the majority of the water has been volatilized.
Water is produced by the dehydrocyclization reaction
and consequently is always present to some extent in
the melt. The imide which has formed by this time is
probably the result of dehydrocyclization of the
succinamic acid and the salt formed by the hydrolysis
reaction. After 15 minutes at 175°C (reaction 8) only
39.5% of the initially present succinamic acid is
accounted for in the residue while the yield of the
dehydrocyclization product, succinimide, is 23.1%. The
yield of ionic ammonia has decreased to 23.1%. Thus,

at longer bake times both the residual succinamic acid
and the hydrolysis product are converted to the imide.

In reactions of succinamic acid the
dehydrocyclization reaction is much faster at higher
temperatures. The yield of imide in reaction 3 at 200°
C is over six times greater than the yield of imide in
reaction 6 at 175°C. The greater competition of the
dehydrocyclization reaction at higher tempertures
accounts quite well for the observations of both
Johnson and Cuculo that the free acid D.S. of rayon
fabric modified in the pad-bake reaction with
succinamic acid decreased at bake temperatures higher
than 170°C.

The formation of succinimide is favored by the
exclusion of water from the bake step. Succinamic acid
which is heated in the absence of water (reaction 22,
melt reaction) yields over five times the succinimide
as does the reaction under normal pad-bake reaction
conditions (reaction 2). The ease of formation of
succinimide in molten succinamic could account for the
low yields in the reactions of succinamic acid melts
with cellulosic substrates (4). Succinamic acid under
the predry conditions (reaction 20) also gave a greater
yield of imide than did the reaction under normal
pad-bake conditions (reaction 2).

Acknowledgements

We wish to thank The Sherwin Williams Co.,
Tennessee Eastman Co., and North Carolina State
University for their support of parts of this work.

List of References

1. Cuculo, J. A., Textile Res. J., 41, 321 (1971).
2. Cuculo, J. A., U. S. Patent 3,555,585 (1971).
3. Johnson, E. H. and Cuculo, J. A., Textile Res. J.
   43, 283 (1973).
4. Allen, T. C. and Cuculo, J. A., ACS Symposium Series
   10, 51, (1975).
5. Cuculo, J. A., Textile Res. J., 41, 375 (1971).
6. Silverstein, R. M. and Bassler, C. D., Spectrometric
   Identification of Organic Compounds, John W. Wiley
   and Sons, New York, 1967, pp. 89-92.
7. O'Connor, R. T., "Analysis of Chemically Modified
   Cotton", Chapter XIII. 4.3. in Cellulose and
   Cellulose Derivatives, Vol. V, Bikales, N. M. and
   Segal, Leon, eds., John W. Wiley and Sons, New York,
   1970, pp. 55-57.
8. Allen, T. C. and Cuculo, J. A., Journal of Polymer
   Science: Macromolecular Reviews 7, 187-262 (1973).

9. March, J. Advanced Organic Chemistry: Reactions,
    Mechanisms, and Structure, McGraw Hill, New York,
    1968, p. 323.
10. Bender, M. L., J. Amer. Chem. Soc. 79, 1258 (1957).
11. Bender, M. L., Chow Y., and Chloupek, F., J. Am.
    Chem. Soc. 80, 5380 (1958).
12. Bruylants, A. and Kezdy, F., Rec. Chem. Prog. 21,
    213 (1960).
13. Bruylants, A., Voortman, B. V., and Crooy, P., Bull.
    Soc. Chim. Belg. 73, 241 (1964).
14. Higuchi, T., Eberson, L., and Herd, A. K., J. Am.
    Chem. Soc. 88, 3805 (1966).
15. Brown, J., Su, S. C. K., and Shafer, J. A., J. Am.
    Chem. Soc. 88, 4468 (1966).
16. Kirby, A. J. and Lancaster, P. W., Biochem. J., 117
    51P (1970).
17. Kirby, A. J., Lancaster, P. W., and Aldersley, M.
    F., Chem. Comm. 1972, 570.
18. Kirby, A. J. and Lancaster, P. W., J. Chem. Soc.
    Perk. 2, 1972, 1206.
19. Rowland, S P., Welch, C. M., Brannon, M. A. F., and
    Gallagher, D. M., Textile Res. J., 37, 933-41 (1967).
20. Rowland, S. P., Welch, C. M., and Brannon, M. A. F.,
    U. S. Patent 3,526,048 (1970).
21. Bullock, A. L., Vail, S. L., and Mach, C. H., U. S.
    Patent 3,294,779 (1966).
22. Gregory, M. J., J. Chem. Soc. Perkin 2, 1972, 1390.
23. Ritter, W., German Patent 526,881 (1928). (C.A. 25
    4893 (1928)).
24. Sauers, C. K., J. Org. Chem., 34, 2275 (1969).
25. Sauers, C. K., Gould, C. L., and Ionannou, E. S.,
    J. Am. Chem. Soc. 94, 8156 (1972).
26. Siegel, E. F. and Moran, M. K., J. Am. Chem. Soc.
    69, 1457-59 (1947).
27. Bowman, B. G., unpublished data.
28. Boltz, D. F., ed., Colorimetric Determination of
    Nonmetals, Interscience, New York, 1958, pp. 84-95.
29. Kolthoff, I. M. and Steizer, V. A., Volumetric
    Analysis II, Interscience, New York 1947, pp. 129-
    30.
30. Niederl, J. B. and Niederl V., Micromethods of
    Quantitative Analysis, John Wiley and Sons, New
    York, 1938.
31. Hargreaves, M. K. and Stevinson, E. A., Spectro-
    chemica Acta 21, 1681-9 (1965).
32. Gregory, M. J. and Loadman, M. J. R., J. Chem. Soc.
    B, 1971, 1862.
33. Antonenko, N. G., J. Gen. Chem. USSR 35, 425-9 (1965).

Modified Cellulosics

Courtaulds Challenge the Cotton Legend

M. LANE and J. A. McCOMBES
Courtaulds Ltd., Coventry, England

In August 1905, Courtaulds Limited began production of viscose rayon in Coventry, England. Six years later, their U.S. subsidiary A.V.C. first made viscose at Marcus Hook, Pennsylvania.

Those two events mark the beginning of over 70 years of Courtaulds' involvement in viscose both in Europe and North America.

70 years in which Courtaulds have remained committed to viscose fibres whilst increasing their involvement in every aspect of the textile trade.

Courtaulds Involvement in Viscose

1905 Production of viscose filament begins in Coventry.
1911 Courtauld's U.S. subsidiary – American Viscose
      Corporation begins production in Pennsylvania.
1926 Introduction of Courtaulds matt filament viscose.
1934 Courtaulds pioneer the introduction of viscose
      staple fibre.
1935 Courtaulds introduce Tenasco, high strength viscose
      tyre yarn.
1957 Introduction of Sarille, chemically crimped viscose.
1964 Courtaulds introduce Evlan, a crimped coarse denier
      carpet fibre.
1965 Vincel 28, polynosic type viscose.
1967 Vincel 64, Courtaulds modal fibre is introduced.
1968 Darelle, Courtaulds flame retardant viscose.
1976 Viloft.

70 years of viscose innovation and development: of crimped fibres, high tenacity and high wet modulus fibres,

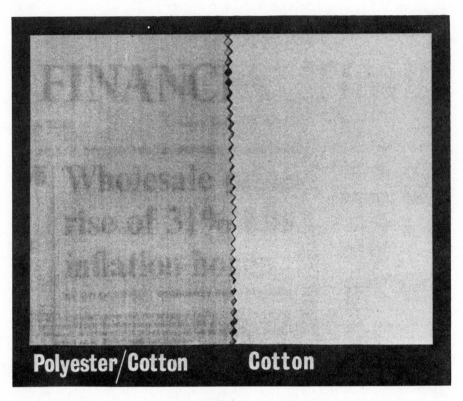

Figure 1.   Fabric cover in sheeting

flame retardants and dye variants, modal fibres, and now
Viloft, a new generation of viscose aimed at a new and an
increasingly aware consumer market.

Viloft Development Background

The development of Viloft begins in the late 'sixties
and early 'seventies, with the growing success of poly-
ester/cotton, and to a lesser extent polyester/Viscose,
in traditional cotton markets.

The benefits of polyester are well known: it brought
durability, wash stability and the easy care concept to
the consumer, and much of our efforts during this period
were aimed at producing and marketing polyester/Viscose
blends in a wide variety of apparel and household textiles.

In the early 1970's, however, we became conscious
of a growing consumer reaction against polyester blends,
with complaints of a synthetic handle and of leanness.
(Figure 1)

Some of these faults were implicit in the use of a
synthetic fibre like polyester, but many were exacerbated
by the use of lower cloth constructions which took advantage of polyester's durability; and by the even higher
proportions of polyester, in blends that moved from 50/50
to 67/33 and onward in order to minimise the influence of
rising cotton prices.

We saw in this situation an opportunity to expand
viscose business if we could produce a high bulk fibre for
blending with polyester. Our aim was to create a blend
which married the wear and care qualities of polyester with
the bulk and cover contributed by a new viscose fibre. To
produce a modern high performance fabric, but with aesthetics more akin to a traditional, natural cotton cloth.

The R & D team to which this challenge was posed
faced only two restrictions.

1.   The new fibre must be produced on existing
viscose machinery and plant.

2.   The new fibre must be processed on conventional
textile machinery, like an existing viscose fibre.

The team's answer was Viloft, a hollow viscose fibre
produced by the inflation process. A fibre which was not
only to meet the bulk requirement of the original project
specification, but which has been found to enhance handle
and comfort in a much wider range of fabrics and blends

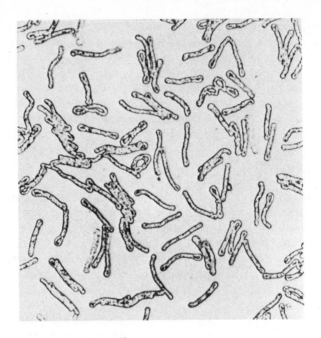

Figure 2. PM type fiber

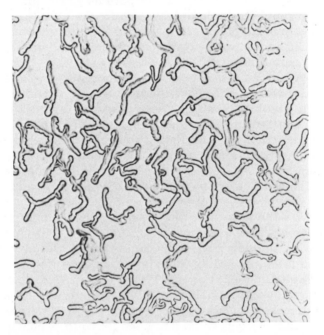

Figure 3. SI type fiber

than originally anticipated.

What is an Inflated Viscose Fibre?

The concept of an inflated viscose fibre is essentially a simple one. Sodium carbonate is added to the viscose prior to spinning. When this carbonated viscose comes into contact with the acid of the spin bath, carbon dioxide gas is evolved.

The art of manufacturing an inflated fibre is to balance the rates of viscose coagulation and regeneration with the evolution of gas, and so contain within the forming fibre that bubble of carbon dioxide.

Inflated viscose fibres in various forms have been proposed for fifty years, but until now no manufacturer has been able to produce commercially a consistent fibre of predetermined cross-section. It is this problem of manufacturing control which Courtaulds, with their long experience of viscose developments, have now overcome.

This ability to control the inflation process opens the way for a whole range of viscose fibres, and the development of this new generation is clearly a major interest in our current viscose research.

There are three fibres of immediate interest. The first, known as PM (Figure 2) is a flat sectioned, self bonding fibre produced by inflating and collapsing the fibre in the spin bath. PM was initially designed for paper making, and in particular for security papers.

Developed from the PM principle is SI fibre - super inflated (Figure 3). Like PM, it is highly inflated, but the spinning conditions are so arranged that its collapse is irregular, forming not a flat, but a multilimbed fibre. SI, a high water imbibition fibre, was produced for surgical and tampon evaluation.

The third fibre is Viloft (Figures 4a, 4b) the first true textile fibre to take advantage of the bulk offered by the inflation process. Viloft is produced by inflating the fibre and fixing it in a hollow tubular configuration.

The cross-sections of polyester and cotton are shown for comparison with Viloft (Figure 5). The similarity of cotton and Viloft is apparent; the dissimilarity of polyester equally clear.

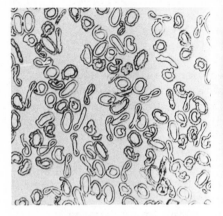
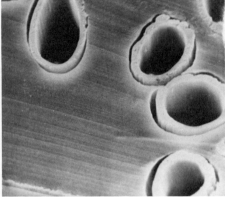

Figures 4a,b. Viloft

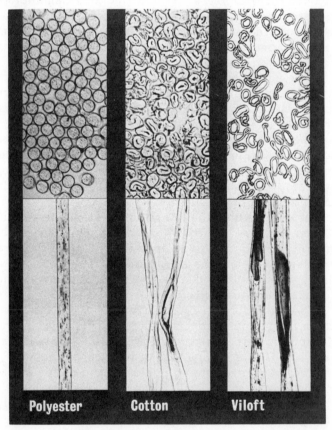

*Figure 5. Cross sections of polyester and cotton compared with
Viloft*

Viloft Manufacture

Viloft is manufactured on standard viscose machinery using process conditions which individually are not unusual, but which in combination give a successful new process. The parameters which are additional to those of the average viscose staple process are the inclusion of sodium carbonate and modifier in the viscose.

The successful manufacture of Viloft depends on two main areas: chemical control of viscose and spin bath conditions to produce a fibre of consistent properties and cross-section, and engineering control to deal with the problems of handling, washing and drying a bulky fibre.

Table I

Courtaulds Viloft Process U.S. Pat 3,626,045

Viscose Conditions		Spin Bath Conditions	
% Cellulose	6.5–9.5	% Sulphuric Acid	8.0–10.0
% Caustic Soda	5.0–7.0	% Zinc Sulphate	1.0– 3.0
% Sodium Carbonate	3.0–4.0	% Sodium Sulphate	20.0–26.0
% Modifier	0.75–2.0	Temperature ^{o}C	25–45
% Carbon Disulphide	33 –50	Hot stretch	2% acid at 95°C
Salt Figure	12 –18		
Ball Fall (18°C)	30 –180 secs.		

Table I shows the chemical conditions required for Courtaulds Viloft. There are six major control areas: salt figure, modifier, and of course carbonate in the viscose making; acid, temperature and sodium sulphate level in the spin bath.

Figures 6 and 7 show how these six parameters separately affect fibre inflation. Each parameter is related to the degree of inflation; each also affects viscose quality, fibre quality and properties.

A change in any one of these conditions alters the balance of the other five, causing the product to alter, in some instances, almost immediately; sometimes over a matter of hours.

It is in this context that the achievement of Courtaulds in producing a consistent inflated product must be measured.

On the engineering front some detailed modifications

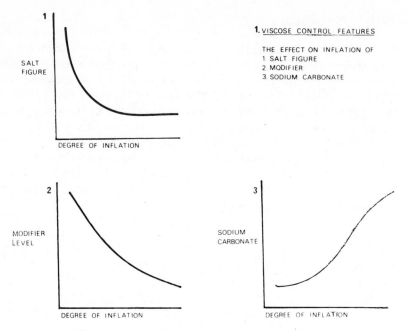

Figure 6.   *Inflated fiber control parameters affecting vicose control*

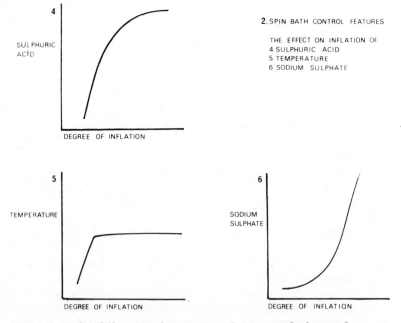

Figure 7.   *Inflated fiber control parameters affecting spin bath control*

have proved necessary. Improved ventilation at the spin bath is required because of the effects consequent on CO_2 release at this stage in the process. Tows of hollow fibre float on the spin bath, being less dense than normal cellulose, and must be controlled, to prevent interaction with other tows causing end breaks. Cutters require modification to cope with a much thicker, bulkier product, and the bed of fibre produced must be washed and dried more efficiently because of its loft and bulk.

Even baling is a problem, and multiple pressings are required to compress the lofty Viloft into a standard weight despatch bale.

The end result of this effort is, we believe, worthwhile.

Fibre Properties

Our original aim in producing Viloft was to make a bulky fibre with similar properties to standard viscose. Table II shows the fibre's physical properties and illustrates that this latter aim was achieved.

Table II

Viloft Tensile Properties

	Viloft	Courtaulds Viscose
Fibre decitex	1.7	1.7
Air Dry Tenacity (cN/tex)	22.0	22.0
Air Dry Extension (%)	14.0	19.0
Wet Tenacity (cN/tex)	11.0	10.0
Wet Extension (%)	18.0	28.0
Loop tenacity (cN/tex)	9.0	8.0
Knot tenacity (cN/tex)	13.0	15.0
Initial wet modulus : 2%	36	30
(cN/tex at 100% extension) : 5%	50	40
Yield point (cN/tex)	7.7	7.3
Yield extension (%)	1.2	2.2

The table shows that in tensile properties Viloft is a close match for standard viscose, though, due to hot stretching, it does have marginally better tenacity and modulus.

The true worth of Viloft is however demonstrated in Table III, which shows the characteristics which make it special.

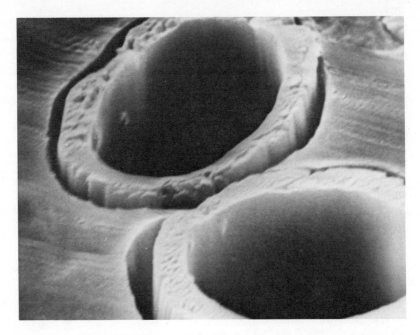

Figure 8.   Viloft—world's first tubular vicose staple fiber in commercial production

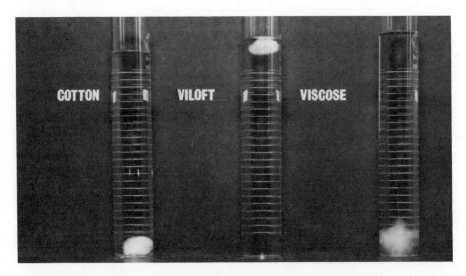

Figure 9.   Measurement of Viloft fiber density

Its high torsional rigidity – a factor contributing to handle.
Its low micronair value – illustrating clearly Viloft's greater
bulk.
Its high water imbibition – related to the fibre's absorbency
and comfort.
Its low effective density – illustrating how much more cover
can be achieved for a given weight of cellulose.

TABLE III

Viloft Special Characteristics

	Viloft	Courtaulds Viscose	American Middling Grade Cotton
Torsional rigidity ($\times 10^{-9}$) ($Nm^2\ tex^{-2}$)	3.7	1.52	3.2
Micronair (cu.ft/hour)	4.7	13.7	8.5
Water Imbibition (%)	130	100	47
Effective Density (g/cc)	1.15*	1.5	1.5

* This is dependent on method of measurement

This last feature – the way in which a hollow fibre reduces apparent density – is shown clearly in Figure 9, in
which fibre densities are being measured, using a flotation
technique.

Processing of Viloft

The processing of Viloft on conventional equipment was
regarded as a cruicial requirement at the inception of the
Viloft project.
    That we can produce a list of successful developments in
Viloft blends, the majority produced under commercial conditions, is a measure of our success, and a tribute not just
to our ability to make a workable textile fibre, but also to the
textile trade's flexibility and its willingness to take and
exploit any new and worthwhile fibre development.
    Our initial concern about the bulk of Viloft causing

problems in yarn spinning has proved groundless. Opener
blends with 50/50 polyester mixtures have been spun on a
commercial scale in counts ranging from 1/14's to 1/40's. In
a 50/50 cotton mixutre yarns from 1/10's to 1/28's have been
produced without difficulty, and open end cotton blend yarns
in coarser counts are now established in the U.K. tufting
trade.

The situation upstream in the textile sequence is equally
encouraging, with good reports on both weaving and knitting
conversion. In dyeing and finishing, Viloft behaves like a
standard viscose, but with the unexpected bonus that with
reactive dyestuffs it dyes closer to cotton. This reduces
partitioning problems in Viloft/cotton mixtures, and extends
the range of dyestuffs available to designer and dyer.

Viloft in Fabrics

The Viloft project specification required a bulky fibre for
polyester blends, and as the fibre properties showed, this
has been achieved. Table IV shows what this additional fibre
bulk offers in fabric. The table compares the air permeability
and light transmission of knitted fabrics produced from 100%
Viloft, Fibro and Vincel, and clearly shows Viloft's advantage
over other rayon types.

TABLE   IV

Fabric Cover 1

A comparison of Viloft and other viscose fibres

	Viloft	Courtaulds Viscose	Vincel
Air Permeability (expressed as air flow in cu.ft/hr)	15.3	19.4	25.3
Light transmission (% light from standard source transmitted through fabric)	4.9	8.1	11.2

Perhaps, however, the most crucial test is shown in
Table V, which compares two dress fabrics - one in 100%
Viloft, the other in 100% cotton - both in identical constructions. Here again Viloft is shown to be superior, its greater
bulk providing a serious challenge to cotton in fabric cover.

TABLE V

Fabric Cover 2

A comparison of fabric cover offered by Viloft and cotton

	100% Viloft	100% Cotton
Air Permeability (cu.ft/hour)	10.8	15.1
Light transmission (%)	1.9	3.7

Even in 67/33, majority polyester blends, where Viloft represents only a minority component, it still contributes significantly to cover. Figure 10 shows three fabrics in equivalent constructions with minority components of Viloft, cotton and viscose. The advantage clearly offered by Viloft is, we believe, significant in consumer terms.

Viloft has not, however, only contributed bulk. It has made an undoubted contribution to the handle and comfort of polyester/Viloft blends, which unfortunately cannot easily be shown in tables or diagrams.

Aesthetics are a matter of critical personal judgment, and we can only ask you to handle and compare Viloft blends for yourself. We are confident, however, that you will agree with so many other textile people that Viloft offers something special.

Certainly, that seems to be the view of more and more cotton users. Viloft, designed to improve the aesthetics of polyester blends, is being used to improve the aesthetics of cotton!

Traditionally viscose was used as a cheapener with cotton: it added little, and could take away much of the attractiveness of the cotton cloth. Viloft reverses that trend. It does not detract from the cover and handle - it complements them - and it brings an improved absorbency, which has proved immediately attractive in diapers and towellings.

The U.K. diaper market is highly conservative and demanding - nothing is too good for baby. Traditionally diapers are white cotton terry squares of full and soft handle, and perhaps the best compliment to Viloft's aesthetics is that 50/50 Viloft/cotton terry warp diapers have been an immediate commercial success. But Viloft does not just feel good: it absorbs more, faster - an important benefit at this end of the market (and that end of the baby). Figure 11 shows that Viloft takes up more moisture than cotton, at a faster rate. We can

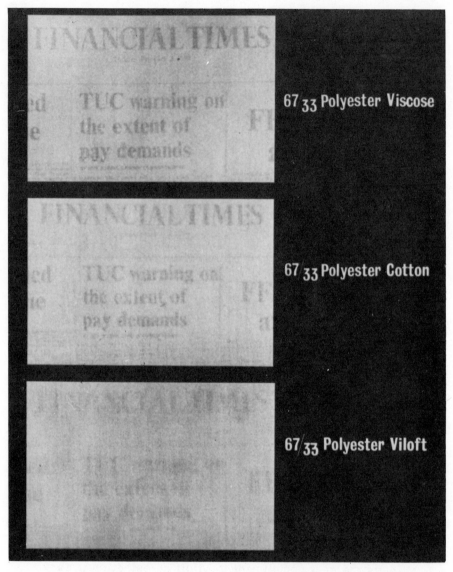

Figure 10.   Comparison of three fabrics in equivalent constructions with minority components of vicose, cotton, and Viloft

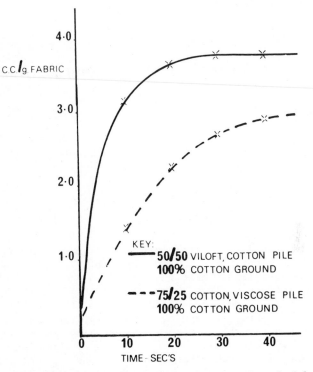

Figure 11. Comparison of moisture uptake on Viloft and standard diapers

also demonstrate that this advantage is preserved in use.

Viloft also offers improved absorbency in towels, but without the drawback of a slimey, sleezy handle, the major limitation which has always prevented the use of viscose in towels. Many laboratory evaluations and user trials have been carried out, with exceedingly promising results, and undoubtedly 50/50 Viloft/cotton terry warp towels will make their debut in the market place in 1977.

Viloft was made as a bulky viscose fibre designed to enhance the aesthetics of polyester/viscose blends. We believe it has become much more than that.

The cover and bulk of Viloft fabrics, their attractive handle and absorbency, add a new dimension to viscose. We have opened up a vast number of possibilities for Viloft, not all of which have been evaluated at this stage in development. Already alternative inflated fibres are being considered, and are competing for a place in the development queue. Prospects look bright indeed.

13

Viscose Rayon Fibers Containing Lignin Derivatives

NEAL E. FRANKS

American Enka Co., Enka, N.C. 28728

The compatibility of cellulose with lignin is demonstrated throughout the plant world. This compatibility is not necessarily shown as an intimate mixture of the two polymers, but rather as a matrix structure of cellulose fibers in lignin. Although chemical pulping systems have been developed to provide dissolving pulps free of lignin, it is surprising that there appear to be no reported attempts to recombine these two polymers in fibrous form. The viscose system could be the ideal technique by which to achieve such fibers, since in the alkaline milieu of the cellulose xanthate solution both polymers will demonstrate similar charge effects. The experiments described here will outline conditions necessary to prepare such fibers and to describe some of the properties of the resulting fibers.

Materials and Methods

The sodium lignate and sodium lignosulfonate used were commercially available materials of highest purity. This degree of purity was deemed necessary due to the possible complications which could arise from the introduction of foreign inorganic materials into the viscose process. Typical analytical values for the sodium lignate and sodium lignosulfonate used are provided in Table 1.

It was necessary to purge the alkaline solutions of the sodium lignate with nitrogen prior to use to remove amounts of ammonia added during the processing of this material. The cross-linking trials of either sodium lignate or sodium lignosulfonate were performed using a 20-25% polymer loading in solution.

Spinning trials were performed by adding alkaline solutions of the lignin derivatives to a standard textile viscose stream just before the last filtration stage. This mixture was spun through a 750 hole spinneret into a rayon spinbath and stretch was applied in a second bath containing hot dilute sulfuric acid. The wet yarn was collected in a centrifugal pot, and the wet yarn cake was purified and dried in normal fashion. The target denier in these trials was 1100 for the dried yarn bundle.

Table 1. Analytical Data for Sodium Lignate and Sodium Lignosulfonate.

	Sodium Lignate (Indulin AT)	Sodium Lignosulfonate (Marasperse CB)
Total sulfur, %	1.6 (as ether)	2.6
Sulfate sulfur as S, %	0	0.1
Sulfite sulfur as S, %	0	0
CaO, %	Nil	0.03
MgO, %	Nil	Trace
Reducing sugars	Nil	0
pH in water	6	8.5-9.2

Results and Discussion

Unmodified Lignate and Lignosulfonate. The fiber denier values obtained by adding an amount of unmodified sodium lignate or sodium lignosulfonate to the viscose stream in amounts to achieve 20 to 40% loadings, based on the amount of cellulose present, are provided in Table 2. The denier values shown are average values obtained using the inside and outside values of the yarn package.

Table 2. Relative Denier Values of Rayon Fibers Spun Using Unmodified Sodium lignate and Lignosulfonate.

Sample	Corrected denier
Control	1110
20% Sodium lignate	1035
40% Sodium lignate	965
Control	1105
20% Lignosulfonate	934

It can be seen that the retention of unmodified sodium lignate was better than that observed for sodium lignosulfonate. This difference is best explained by the fact that the sulfonic acid residues present in the sodium lignosulfonate cause this polymer to be more soluble in the subsequent purification stages of the viscose rayon process allowing a greater loss of lignosulfonate. It was felt that some technique was needed to improve the retention of both sodium lignate and lignosulfonate; a method of cross-linking was judged the best possibility of achieving this end.

Cross-linking Methods. The choice of the proper cross-linking method was based on the ease and reproducibility of the technique available. Oxidative cross-linking, such as that reported by Nimz[1], was eliminated on the basis that it could exascerbate the problem of color in the resulting fiber. A second cross-linking method considered was that of etherifying the reactive phenolic hydroxyl residues with multifunctional cross-linking agents such as epichlorohydron or cyanuric chloride[2]. Small scale experiments demonstrated that this method was not easy to control. Further, a significant reduction of the phenolic hydroxyl population could adversely affect the solubility of the lignin derivatives, particularly sodium lignate, in the viscose system.

The method finally adopted for cross-linking both the sodium lignate and lignosulfonate was the use of formaldehyde under alkaline conditions. Pilot experiments were performed by cross-linking at pH 10.5-11; a crude measure of success was obtained by pouring these cross-linked materials into spin-bath to observe the yield of cross-linked material. Optimum formaldehyde levels were found to be 5% w/w for sodium lignate and 10% w/w for sodium lignosulfonate. This difference is a further reflection of the solubility difference due to the sulfonic acid residues present in sodium lignosulfonate.

Spinning trials were again performed, but using the cross-linked derivatives. The results obtained with the addition of 20% and 40% levels of formaldehyde cross-linked sodium lignate are provided in Table 3. It is likely that some error occurred in the determination of the solids content of the cross-linked sodium lignate, which was reflected in the lower fiber denier determination. As expected, addition of a diluent to the cellulosic fiber reduced both the conditioned and wet tenacities as well as reducing the elongation at break.

A denier error in the opposite direction was observed for the fiber containing 20% formaldehyde cross-linked sodium lignosulfonate (Table 4).

Table 3. Physical Properties of Rayon Fibers Containing Formaldehyde Cross-Linked Sodium Lignate.

	Control	20%	40%
Denier	1110	1075	1018
Cond. ten. (gpd)	2.6	2.16	1.94
Cond. elong. (%)	24.4	21.8	19.6
Wet ten. (gpd)	1.3	1.03	0.82
Wet elong. (%)	37.0	34.9	30.4

This difference is again attributed to the difficulty of obtaining an accurate total solids content of the cross-linked mixture. The decrease in the other physical properties closely matches those observed for the cross-linked sodium lignate system.

Table 4. Physical Properties of Rayon Fiber Containing
Formaldehyde Cross-Linked Sodium Lignosulfonate.

	Control	20%
Denier	1105	1180
Cond. ten. (gpd)	2.67	2.22
Cond. elong. (%)	26.2	19.9
Wet ten. (gpd)	1.35	0.96
Wet elong. (%)	37.0	28.5

Demonstration of Included Lignin Derivatives. It was
desirable to have a method demonstrating the presence of the
cross-linked lignin derivatives, in addition to the denier
determinations used in the early portion of this discussion.
Attempts were made to demonstrate the presence of these added
polymers chemically; iodination and measurement of the level of
iodination of the phenolic residues was inconclusive.
Selective extraction of the cellulose portion of the fiber with
cupriethylenediamine dissolved both the cellulosic and lignin
portions of the fiber.

An examination of some additional physical properties
(Table 5) showed some slight differences, but nothing of
sufficient magnitude to convincingly demonstrate the presence
of the added polymers. The slight decrease in water imbibition
of the cross-linked sodium lignate fiber was in the expected
direction since the lignate would be expected to have some
hydrophobic character. Likewise, the cross-linked sodium
lignosulfonate had an improved water imbibition. Neither of
these values were far enough from control values to be con-
vincing. The same comments can be made for the X-ray order
and knot tenacity retention values.

Table 5. Water Imbibition, X-Ray Order, and Knot Tenacity
Retention Values of Rayon Fibers Containing
Cross-Linked Lignins.

Sample	Water Imbibition,%	X-Ray Order	Knot Tenacity Retention, %
Control	80-90	0.87	79
20% Cross-linked lignate	72	0.85	77
20% Cross-linked lignosulfonate	96	0.85	71

Considerably more success was experienced using infrared
difference spectroscopy in the 5.75-6.25 micron region. These
spectral studies were performed using both control fibers and
sample fibers in KBr pellets. Calibration curves were obtained
by mixing known amounts of the cross-linked lignin derivative
and control rayon fiber. By this measure, the rayon fiber

thought to contain 20% cross-linked sodium lignate was shown to contain but 18% of the included polymer. This explains the lower than expected (Table 3) denier values for this fiber.

Color. It will be appreciated by those familar with the pulp and paper industry that addition of lignin derivaties to rayon fibers creates a fiber that has a distinct reddish-brown hue. Quantitation of this color using the Hunter L A B method is shown in Table 6. In this system, the L value is a measure of lighness, the A value indicates a red hue if positive and a green hue if negative, and the B value indicates yellowness if positive and a blue tint if negative. By this measure, the fiber containing the formaldehyde cross-linked sodium lignate is darker than the corresponding fiber containing cross-linked lignosulfonate, although it has a yellowness value. The values observed, when compared with a normal control fiber, would imply that the fibers containing the cross-linked lignin derivatives are unacceptable for normal textile uses.

Table 6.   Hunter L A B Values for Rayon Fibers Containing Cross-Linked Lignin Derivatives.

Sample	L	A	B
Control	85.0	-3	+5
20% Cross-linked lignate	34.0	+6.8	+15.1
20% Cross-linked lignosulfonate	47.5	+6.4	+19.8

Conclusion

It is possible to spin regenerated cellulosic fibers containing up to 40% formaldehyde cross-linked sodium lignate or lignosulfonate using the viscose process. There is a corresponding decrease in tenacities and elongation at break for such fibers. The presence of the added polymers was best shown using infrared difference spectroscopy. There is a considerable increase in the color of the rayon fibers containing these derivatives due to the chromoporic residues present in the included polymers.

Literature Cited

1. Nimz, H., German Patent 2,221,353.
2. Allan, G. G., U. S. Patent 3,600,308.

A Process for Drying a Superabsorbent Pulp

P. LEPOUTRE

Pulp and Paper Research Institute of Canada

The market for bleached pulp in absorbent sanitary products is rapidly expanding. Pulp, generally in fluffed form, is quite efficient for absorbing quickly fairly large quantities of liquid. Conventional pulp has, however, serious drawbacks. One is the bulk of the products, which require large storage space. Another is the relative ease with which a large proportion of the absorbed liquid is released under light pressure.

There is a need for an absorbent material, preferably in fibrous form, capable of absorbing larger volumes of liquid per gram of absorbent and to hold this liquid under moderate pressure. Several such "superabsorbent" products, some in powder form, some in fibrous form, have been developed in the last few years (1,2).

A superabsorbent in fibrous form has been developed at the Pulp and Paper Research Institute of Canada to the completion of the laboratory stage; its preparation and properties have been the subject of several publications (3,4,5). Essentially, it consists of wood pulp modified by graft-polymerization of polyacrylonitrile which is subsequently hydrolyzed into a copolymer of sodium polyacrylate and polyacrylamide which confers on the pulp an outstanding affinity for water.

In addition to its water absorption properties, this product, or variants of it, may have considerable interest in wet and dry forming of specialty papers.

Commercialization of such a material hinges upon the technical and economic feasibility of producing it in dry form. Conventional thermal drying is slow, causes the fibers to lose some of their absorptive capacity and furthermore produces a stiff and brittle sheet which is impossible to fluff. Freeze-drying or solvent exchange drying prove much too costly because of the huge quantity of water to be removed.

An unconventional drying method had to be developed; this report deals with the development of this novel drying process.

I. Requirements to be met

The requirements that had to be met are listed below:
1. The drying process should not impair the absorption
   properties and should leave the product in fibrous form.
2. The dry product should be fluffable by ordinary means.
3. The product should be economical as judged by its cost/
   performance ratio when compared to regular pulp.

II. First Attempts

The first attempts to produce a product meeting these requirements will be described as they point to a step critical
for success.
   1. Hydrolysis under non-swelling conditions. Under normal
conditions the grafted polyacrylonitrile was hydrolyzed in
boiling aqueous NaOH. The pulp was then filtered and washed to
remove excess NaOH. Typically, the quantity of liquid associated
with the fibers was of the order of 8-10 g/g fiber after hydrolysis and of 30-40 g/g after washing, as compared to approximately
1 g/g for the untreated fibers.
   To avoid the large quantity of water taken up during hydrolysis in aqueous NaOH, it was logical to attempt the hydrolysis
under low-swelling conditions. Hydrolysis experiments carried
out in the presence of alcohols showed that this could be
achieved. In this fashion the amount of liquid retained by the
fibers at the end of hydrolysis was considerably decreased, and
further solvent exchange drying was easy. However, this route
had disadvantages:
   a) hydrolysis at reflux temperature was quite long and
      sometimes incomplete unless substantial amounts of water
      were present, depending on the alcohol used.
   b) while the water absorption characteristics of the dried
      fibers were usually close to normal, the absorption of
      1% aqueous NaCl (simulating physiological fluids) was
      quite low and proportional to the amount of water pre-
      sent in the hydrolysis liquor.
   2. Hydrolysis at high pulp concentration. Another way to
reduce the amount of liquid taken up is to reduce the amount of
available liquid. Hydrolysis experiments with aqueous NaOH at
high fiber concentration followed by solvent-exchange drying,
produced fibers which swelled normally in water but not in NaCl.
Swelling in aqueous NaCl was inversely proportional to the fiber
concentration used, i.e., proportional to the amount of water
present during hydrolysis.
   3. Normal aqueous hydrolysis followed by solvent-exchange
drying. The preceding experiments seemed to indicate that the
swelling taking place during normal, low consistency, aqueous
hydrolysis was necessary. Since most of the total swelling takes
place in the washing stage (liquid associated with the fibers

increases from 8-10 g/g to 30-40 g/g), solvent exchange drying
was done immediately after hydrolysis. Again, while the water
absorbency was normal, the salt absorbency, while higher than in
the previous experiments, was still low (7-10 g/g), but became
normal after the fibers had been soaked in water prior to con-
tacting them with the salt solutions.

III. Stage of full swelling prior to drying: a requisite

During the course of this work it became evident that whenever fibers were dried without having undergone full swelling in
water at pH 6-9 in the absence of a significant concentration of
salt, they would reswell in water more or less to their normal
extent but not in a salt solution.

Many experiments then confirmed that taking the fibers
through their state of full swelling prior to drying was a requisite for salt solution absorption.

In a previous article (3) it was explained that swelling
was due to osmotic pressure effects and that equilibrium was
reached when a balance was achieved between these swelling forces
and the fiber cohesion forces. While alkaline aqueous hydrolysis
contributes to a substantial reduction in the fiber cohesion, considerably more weakening takes place during the swelling that
occurs during the washing operation and the transformation is ir-
reversible, as seen on Fig. 1, line ABC.

This requirement of a full swelling stage during the manufacturing process meant that a new approach had to be thought
of.

IV. The drying process

1. Principle. The process makes use of the solution properties of the grafted polymer. This polymer is soluble in
water at pHs above 4 but insoluble at lower pHs and in alcohols.
As seen in Fig. 2, swelling as measured by the amount of water
retained after high speed centrifugation (water retention value,
WRV) decreases from 30 at pH 6-9 to 2.5 g/g when the pH is reduced to 3. At this pH, the grafted polymer is in its non-
ionized polyacrylic acid form. If the fibers are dried at that
point, they will not reswell in water to any extent. However if
the deswollen fibers are converted back into their sodium salt
form with NaOH under non-swelling solvent conditions before
drying them, normal reswelling in water and in salt solution will
take place.

The process then consists of three steps.
a. The hydrolyzed fibers are taken through their maximum
swelling state at pH 6-9 in the absence of any significant salt
concentration.

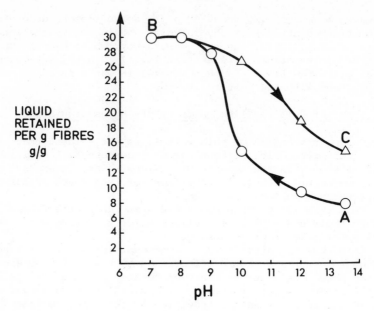

Figure 1. Liquid retained by the fibers as a function of the pH

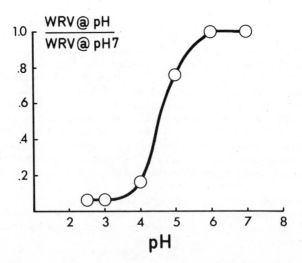

Figure 2. Relative water retention value as a function of the pH

b.   The pH is brought down to approximately 3 where the fibers are in their minimum swelling state.

c.   The acid groups are converted into sodium carboxylate groups with NaOH in alcohol (a non-solvent for the polymer) and the alcohol is evaporated.

2. Laboratory procedure. At the end of hydrolysis the fibers are filtered, washed with one volume of water while on the filter. The pH at that point is 9.5 - 10. The fibers are then slurried in water at 1% consistency and sulfuric acid is added with good agitation to avoid any local high concentration. During the acidification, the fibers undergo full swelling in the pH range 8-6 in the absence of any significant salt concentration. As the pH continues to drop, the fibers start to de-swell as evidenced by a rapid decrease in slurry viscosity. At pH 4 flocculation starts and acidification is continued until pH 3. At that point, the fibers are in their minimum swelling state, and are flocculated in large aggregates. They are soft and sticky and will form strong wet bonds when pressed together. The slurry is then drained freely or filtered with gentle suction until water starts to recede from the fibrous pad surface. Suction is released. This operation is critical when a fluffable pulp is desired. Excessive compaction of the pad must be avoided because of the tendency of the soft fibers to form strong interfiber bonds which slows down further processing, and makes fluffing very difficult.

While the fibrous pad is still on the filter, a solution of NaOH in aqueous methanol (e.g., NaOH/H_2O/MeOH - 1/9/90) is poured on the pad to displace the interfiber capillary water which is allowed to drain freely or under gentle suction. NaOH converts the carboxylic acid groups into sodium carboxylate while MeOH prevents swelling taking place. Once all the inter-fiber capillary water has been replaced by the methanolic sol-ution the fibrous mat becomes stiff because the polymer, which is insoluble in methanol, "precipitates". From there on, strong suction can be applied. A solution of aqueous methanol (e.g. H_2O/MeOH-8/92) is poured to wash out the excess residual NaOH, followed by pure methanol to remove the last traces of water. The methanol is then evaporated to yield a bulky mat of loosely held fibers.

3. Suggested industrial drying process. Two alternative drying process schemes are offered for consideration. The first one is continuous; the second is a batch progress.

a. As shown schematically in Fig. 3, the acidified fiber slurry is delivered through a headbox on an endless screen. Water drains freely in the first section. This stream is dis-carded. Alkaline aqueous methanol is sprayed on top of the web while gentle suction is applied. Once the solution has dis-placed the water, strong suction is applied to remove most of the excess liquid, from which methanol and NaOH may be recovered in a separate recovery cycle. The web is then sprayed with NaOH-

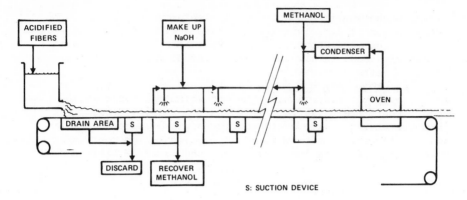

Figure 3.   *Continuous drying process*

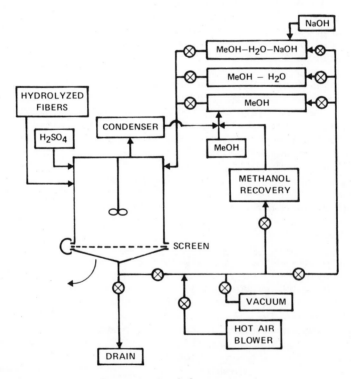

Figure 4.   *Batch drying process*

free, aqueous methanol to remove excess NaOH and continue the
dehydration process. It passes over another strong suction device to remove the excess liquid which is sent to the first
treatment zone after NaOH make-up. The web then receives a spray
of pure methanol, passes over another suction device and goes to
an oven where methanol is evaporated and recycled.

It should be understood that dehydration of the fibers takes
place continuously throughout the sequence. The solvent solution
runs counter-current to the web and its concentration is adjusted
at each treatment zone so that, at no point, do the fibers pick
up water from the solvent solution. In the first section, dehydration is favored when both the concentrations of NaOH and
MeOH are increased. However, increased NaOH concentration means
a longer washing zone. In the second section, dehydration is
favored by higher MeOH concentration but removal of excess NaOH
is slowed down. (The equilibrium isotherm between concentration
of water in the solvent and absorbed water in the fiber, though
not available now, will be required for this control).

b. The alternative process is a batch operation. It is
really a scaled-up version of the laboratory procedure and is
shown in Fig. 4. It has the disadvantage of being discontinuous
but would seem to be much easier to control and be lower in
capital cost. The acidification tank is equipped with a rotating
bottom fitted with a screen. The washed hydrolyzed fibers are
fed to the acidification tank and H_2SO_4 is added to bring the pH
down to 3.0. Then the free draining liquid is removed from the
slurry through the screen until water starts to recede from the
interfiber capillaries. The filtrate is discarded. At that
point, a quantity of alkaline methanolic solution equivalent to
at least one pad volume is poured on top of the fiber pad. Its
flow-rate is adjusted so that friction caused by the flow is
minimized to avoid compaction of the pad and the alkaline methanolic solution has had time to diffuse into the fiber flocs.
When NaOH - depleted aqueous methanol arrives at the bottom, the
drain valve is closed, and the filtrate is sent for recovery of
methanol. One volume of aqueous MeOH solution is run through
the pad to wash out the excess NaOH followed by one volume of
methanol. The filtrates, depending on their composition, are
recycled to the storage tanks.

After the pure methanol wash, suction is
applied through the pad by the vacuum source to remove most of
the methanol. Then hot air is blown up through the fibrous cake
and methanol is condensed. Finally, the dry cake is discharged
through the revolving bottom, shredded and baled.

The process has been sucessfully proven in the lab on
fiber cakes 2 feet high and 2 inches in diameter.

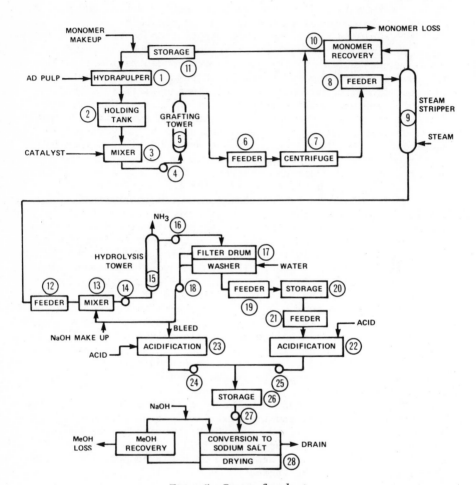

Figure 5.   Process flowsheet

V. Flow-sheet and Mass Balance for the Complete Process

A. Flow-sheet

Fig. 5 shows the flow-sheet for the complete production process from grafting through to drying.

1. Grafting of polyacrylonitrile. Air-dried pulp is slurried at 4.35% consistency using recycled streams (water, HNO_3, acrylonitrile) in a hydrapulper or similar equipment equipped for handling acrylonitrile (toxic) solution and acid resistant to pH 2.5. The slurry is pumped to a holding tank. (In an integrated mill operation, wet lap pulp would be used and no hydrapulper is necessary).

The grafting reaction is done in a continuous fashion. From the holding tank the slurry is sent to a small (residence time ~ 30 sec) mixer where the catalyst is introduced. Consistency is now 4%. The slurry is pumped to a "grafting tower" where no further mixing takes place. Laboratory adiabatic grafting experiments have shown that polymerization is complete after 3-4 min., suggesting a residence time of 5 min.

The very bulky, free-draining grafted pulp is then fed to a centrifuge (residence time 1 min.) and the liquid containing the unreacted monomer is recycled to the hydrapulper.

The centrifuged grafted pulp is then steam-stripped of its residual monomer which is recovered. Residence time: 15 min.

2. Hydrolysis. After the steam-stripping operation, the pulp is slurried in a small mixer at 2% consistency in the recycled hydrolyzing liquor (residence time: 30 sec.) and pumped to a hydrolysis tower equipped with steam jacket. Residence time is 15 min.

From the hydrolysis tower the slurry is pumped to a drum filter where it is dewatered and washed. Provision is made to separate the filtrate and washings to minimize NaOH carry over in the bleed. (See mass balance for reasons for this bleed). This bleed is acidified to recover the dissolved polymer. The bleed and the fibers are acidified separately since the substantial quantity of Na_2SO_4 produced in the acidification of the bleed would be detrimental to the full swelling of the fibers.

The washed fibers, as a swollen fibrous mat, are best handled by a screw-type extruder, and sent to storage.

3. Acidification and drying. The next step, acidification, is carried out batch-wise at a consistency of 1.5%. The acidified slurry is then processed as described in IV, 3. A special type of pump, possibly a plug-flow piston pump, is required to avoid the compaction of the soft, sticky fiber aggregates. Gravity flow should be utilized as much as possible.

$$\text{Cell} - \begin{bmatrix} CH_2 - CH - CH_2 - CH \\ \qquad | \qquad\qquad | \\ \qquad CN \qquad\qquad CN \end{bmatrix}_n + n\ NaOH + 2nH_2O \rightarrow$$

$$\text{Cell} - \begin{bmatrix} CH_2 - CH - CH_2 - CH \\ \qquad | \qquad\qquad | \\ \qquad CO \qquad\qquad CO \\ \qquad | \qquad\qquad | \\ \qquad NH_2 \qquad\quad ONa \end{bmatrix}_n + n\ NH_3$$

Yield, as experimentally determined, was 1.53 part of hydrolyzed copolymer per part of PAN. This corresponds to ca. 50/50 w/w ratio of amide to acid groups.

Hydrolysis causes a loss in both cellulose and polymer: a) Cellulose loss has been set at 5% (experimental finding) and is not recovered. It is assumed to leave with the liquid drained after acidification. b) Polymer loss during hydrolysis has been set at 13% (experimental finding). However the dissolved polymer may be recovered by bleeding enough of the washings to maintain the concentration of dissolved polymer in the hydrolyzing liquor at approximately 2%., acidifying the bleed and adding the precipitated polymer to the acidified fibers.

3. Washing. The quantity of water used to wash the hydrolyzed, filtered fibers has been set as equivalent to the quantity of liquid carried over with these fibers. Laboratory experiments using drop-wise addition of that amount of water to the filter cake under suction showed that the amount of NaOH left with the fibers was sufficiently low that the Na_2SO_4 produced during subsequent acidification did not prevent full swelling of the fibers. The pH of the washed fibers is approximately 10.

4. Acidification. Since the fully swollen fibers are themselves about 3% solids, the provision of adequate free water for mixing and homogeneous acidification requires a lower consistency and 1.5% has been chosen.

Chemical reactions:

$$\text{Cell} - CH_2 - \begin{bmatrix} CH - CH_2 - CH \\ | \qquad\qquad | \\ CO \qquad\quad CO \\ | \qquad\qquad | \\ NH_2 \qquad\quad ONa \end{bmatrix}_n + \frac{n}{2} H_2SO_4 \rightarrow$$

$$\text{Cell} - CH_2 - \begin{bmatrix} CH - CH_2 - CH \\ | \qquad\qquad | \\ CO \qquad\quad CO \\ | \qquad\qquad | \\ NH_2 \qquad\quad OH \end{bmatrix}_n + \frac{n}{2} H_2O + \frac{n}{2} Na_2SO_4$$

$$2NaOH + H_2SO_4 \rightarrow Na_2SO_4 + 2H_2O$$

B. Mass Balance

The mass balance is shown in Fig. 6 and the conditions used are based on laboratory results.

A PAN graft level of 120% was chosen for the mass balance calculations used in estimating the cost, because the absorption properties measured on undried fibers level off at this point (3). Recent data have indicated that, in the case of fibers dried by this process, the leveling off effect occurs at a somewhat higher graft level (150-160%). Typically the WRV and SRV of fibers at 120% PAN graft level are 30 and 13.5 g/g respectively while at 150% the WRV is in excess of 30 and the SRV is 16 g/g.

1. Grafting. Grafting is performed using the ceric ammonium nitrate initiation method.

Catalyst requirement was found to be independent of pulp concentration when the latter was 3% or greater (5), so that it is more economical to operate at as high a pulp concentration as possible. However, the grafting reaction is very fast and a concentration of 4% has been chosen in order to have a rapid homogeneous mixing of the catalyst. A graft level of 120% can be achieved by various combinations of catalyst and monomer concentrations, (3). Increasing catalyst concentration increases grafting cost but leads to higher conversion of monomer to polymer thereby reducing monomer recovery cost. The conditions used (catalyst 1.3% of pulp, monomer concentration 9%) are in the range where cost is optimized. The fate of the ceric ion is still undetermined, and it has been left out of the mass balance.

2. Hydrolysis. The concentration of NaOH in the hydrolysing liquor has been set at 3% w/w. Hydrolysis rate is enhanced by higher NaOH concentration, so that the residence time, and thereby the hydrolysis tower size, could be reduced. However, higher NaOH concentration causes:

  a. increased pulp degradation and loss.
  b. increased residual NaOH at the washer, thereby slowing down the operation.

A consistency of 2% has been chosen because the hydrolyzed pulp is already significantly swollen. It might however be possible to operate at 3% consistency, which could permit a smaller tower size.

Chemical reactions during hydrolysis:

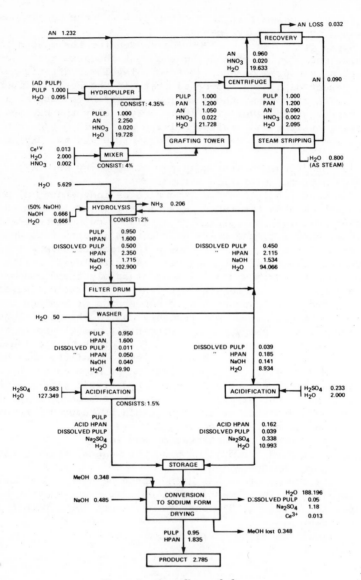

Figure 6. Overall mass balance

5. Conversion to Sodium Salt and Drying by Methanol Exchange. No detailed balance is made for this operation as no data are available for counter-current continuous operation.

It is assumed, again based on static laboratory experiments, that approximately 7 parts of methanol are required per part of product, and are recovered by distillation.

0.125 part of MeOH per part of product (250 lbs/ton product) is assumed to be lost in the bottoms of the distillation tower and in leaks. This represents approximately 2% of the methanol flow.

VI. Modification of Process for Improved Saline Retention Value

It has been explained earlier that swelling is restricted by the fiber cohesion forces. If this cohesion is weakened by some means, e.g. by beating, the water and sodium chloride retention values may be improved by as much as 30%. A saline retention value of 20 g/g has been achieved.

The beating may be performed either on the pulp prior to grafting or on the hydrolyzed pulp.

a. Beating the pulp prior to grafting. In addition to increased absorbency as shown in Fig. 7, beating of a dried pulp leads to higher grafting efficiency (5).

Care should be taken not to overbeat the pulp as this would create serious problems when filtering and washing the hydrolyzed product.

b. Beating the hydrolyzed pulp. This is best achieved by submitting the fragile swollen fibers to violent mixing before the acidification step. Here again, care should be taken not to overbeat the fibers as they could be transformed eventually into a very thick colloidal solution (4). An example of the improvement to be expected is shown in the following Table I.

TABLE I.
Improvement in SRV by Shearing the Hydrolyzed Fibers

Initial SRV	SRV after stirring at high speed for	
g/g	5 min.	20 min.
13.5	16	18.5*
16.6	19.8	22*
18	24*	

* Fibers begin to lose their fibrous character.

VII. Other Products

1. Non-hydrolyzed PAN Absorbent Pulp. The intermediate product, unhydrolyzed PAN grafted pulp, has quite interesting properties which warrant its consideration as an absorbent

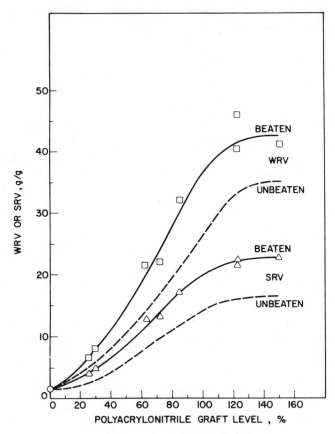

Figure 7. Influence of beating on WRV or SRV

material as such. It may be compared to cross-linked cellulose
fibers.

Cross-linked cellulose fibers in mat or fluff form are
superior to regular pulp in rate of absorption of fluids and in
total capacity, particularly when absorption takes place under
moderate pressure. This is because the cross-linked fibers have
a higher wet modulus so that the mat or fluff does not collapse
when absorbing fluids under moderate pressure. In addition,
little or no interfiber bonding takes place during drying so
that the comminution of the pulp is easily accomplished. Several
patents have been issued in this area.

Pulp fibers grafted with PAN, a glassy, water-insensitive
polymer, behave quite similarly to cross-linked cellulose fibers.
They are bulky, stiff when wet, and do not form interfiber bonds
during drying. For the same reasons as stated above, PAN
grafted pulp is superior to regular pulp in fluid absorbency
under moderate pressure, as shown in Table II. Furthermore, the
absorption being of a capillary nature, it is not affected by
the ionic strength of the fluid.

TABLE II

Pulp	1% NaCl absorption capacity, g/g	
	under 1.7 KPa	under 6.9 KPa
Bleached Spruce Kraft A	12	8
Bleached Soft Kraft B	13	9
Pulp B grafted with PAN	18	13

A plant producing the superabsorbent pulp could also supply
the cheaper PAN graft intermediate for applications where intra-
fiber absorbency is not required.

2. Superabsorbent Pulp in Sheet Form. In some applications, the superabsorbent pulp may be used in sheet form. Then,
fluffability is no longer a requirement, but good sheet formation
becomes important.

As indicated before, acidification to pH 3.5 - 3.0 causes
flocculation of the fibers into aggregates and this results, of
course, in poor sheet formation. However, a further decrease of
the pH to approximately 2.2 will create ionization of the amide
groups and redispersion of the flocs. Very good sheet formation
is then obtained.

Since fluffability is no longer a requirement, the sheet
can be compacted and suction can be applied immediately after
sheet formation. This permits considerable reduction in the
quantity of methanol to be recovered by distillation.

   3. Application in Dry Forming. Preliminary laboratory experiments have demonstrated that fibers prepared by this process
and dry-laid, can form sheets with good interfiber bonding when
pressed as ca 80% RH. The moisture content at this RH is sufficient to plasticize the fibers and to lead to the development of
interfiber bonds. These bonds however are not water resistant.
   More attractive is the possibility of utilizing solvent-exchange dried fibers in the acid form (no conversion to sodium
carboxylate). Here, the interfiber bonds exhibit temporary wet
strength. Considerably more work is required to confirm their
potential as self bonding fibers in dry-forming processes.

ABSTRACT

   A process for the production of a fibrous superabsorbent is
described. The superabsorbent is obtained by grafting polyacrylonitrile on a bleached pulp, hydrolysing it to a copolymer
of sodium polyacrylate and polyacrylamide and removing the water
from the swollen fibers by a combination of chemical dehydration
and solvent-exchange drying. Fibers are obtained in an easily
fluffable form and have water and 1% aqueous NaCl retention
values of up to 40 g/g and 20 g/g respectively. Process flow-
sheets and mass balance are given.

Literature Cited

1. Chemical Week, (1974) July 24, p. 21
2. U.S. Pat. 3,589,364
3. Lepoutre, P., Hui, S.H. and Robertson, A.A. J. Apply.
    Polym. Sci. (1973) 17, 3143.
4. Lepoutre, P. and Robertson, A.A. Tappi, (1974) 57 (10), 87
5. Lepoutre, P. and Hui, S.H., J. Appl. Polym. Sci. (1975)
    19, 1257.

Use of Fiber and Chemical Microscopy in Applied Woven and Nonwoven Technology

ARTHUR DRELICH

Absorbent Technology, Johnson & Johnson, c/o Ethicon, Inc., Somerville, N.J. 08876

DAVID ONEY

Research Div., Chicopee Manuf. Co., Milltown, N.J. 08850

Chemists traditionally use structural formulas and mathematical expressions to understand and to figuratively see technical problems and scientific phenomena. There are many technical areas where it is valuable to literally see materials and the changes they undergo as a result of chemical or physical processes. Understanding through seeing is especially useful in fiber and textile technologies. This seeing is carried out using optical and electron microscopy. The experimenter must understand the capabilities and the limitations of his equipment, and develop a good level of competence in using it. Otherwise, his visual evidence will be, at best, incomplete and at worst, unreliable or even incorrect. In practice he must also develop adequate skills in scientific photography.

The modern history of textile microscopy started with a book by Prof. Schwarz (MIT), Textiles & the Microscope (1934). Other important early contributors include George Royer and Charles Maresh of Cyanamid, and Mary Rollins of the Southern Regional Lab of the USDA.

Woven Fabrics

In defining woven fabrics, two of the most important parameters are the geometry of the yarns in the overall structure, and of the fibers within the yarns. Figure 1 shows an 88 x 88 count cotton fabric, original magnification of 20X, illuminated from below by the bright field method. The information obtainable from this picture includes thread

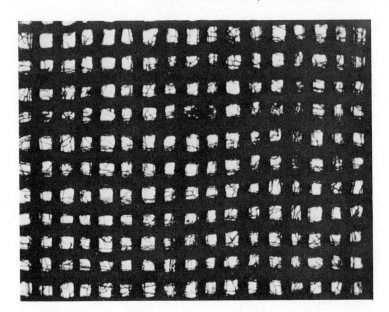

Figure 1

Figure 2

count, uniformity of weave and of yarn thickness, and dimensions of open areas.

Figure 2 shows the same area of this fabric illuminated from above by incident lights at an angle about 45°. Here we can see the surface texture or roughness, and an indication of fiber structure. This is the usual way that fabrics are pictured.

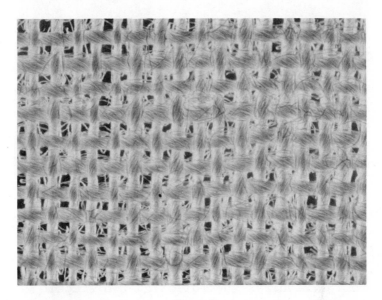

Figure 3

Figure 3, still showing the same area of the same fabric, highlights the fibers which make up the yarns. Individual fibers are clearly resolved, even though their actual diameters are only about 15 micrometers (microns). The direction and angle of twist can readily be measured.

Figure 4 shows a higher magnification (of #3) for ease of seeing detail. Literally every surface fiber of every yarn is clearly visible.

The following two Figures (#5 and #6) are of fabrics with more complex weaves. The precise weave and spinning structures are visible and can be quantitatively measured.

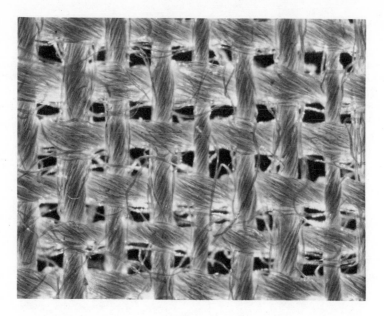

Figure 4

The technique for making the surface fibers so clearly visible (as in Figures 2 through 6, and Figure 10) requires the following two steps:

1. The fiber structure is lightly coated with gold or other suitable metal in a vacuum evaporator or DC sputterer, and

2. The specimen is illuminated in a substage darkfield mode for viewing and photography.

Cross-sectioning of fabrics is now well known. Some of the pioneer work in this field was done by George Royer and others of American Cyanamid from about 1942 to 1950, and published in the Textile Research Journal and the Journal of the Society of Dyers and Colorists. They demonstrated the dyeing and finishing of fabrics by penetration of these reagents into yarns and fibers using cross-sections which were treated by differential staining techniques. These were patterned after the methods biologists use to show tissue structure.

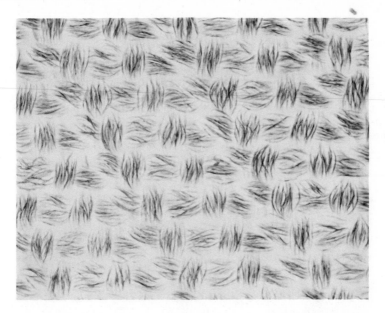

Figure 5

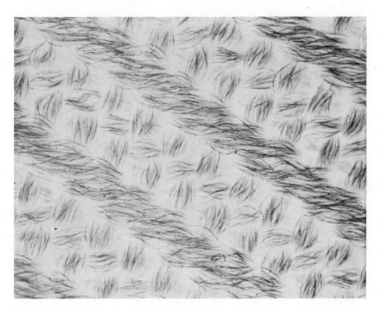

Figure 6

A further method not often used is the planar
sectioning of fabrics to reveal and define interior
structure, where the sections are cut parallel to
the major plane of the fabric. An example is shown
in Figure 7. This is a composite structure of two
woven fabrics bonded together by fused nylon powder
at the interface. This photomacrograph is one of a
series showing the depth of penetration of the fused
powder into the two fabrics. This photograph was
taken by polarized light which acted to effectively
differentially stain the fibers and adhesive.

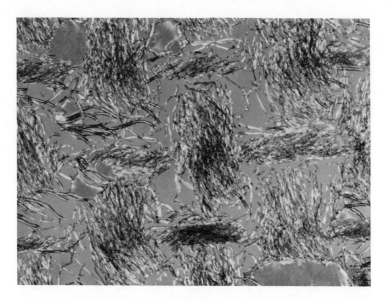

Figure 7

Non-Woven Fabrics

The structure of non-woven fabrics defines, to a
considerable extent, their properties. The distribution and orientation of the fibers can explain
relative long and cross tensile properties (Figure 8,
an oriented card web). A photomacrograph of a non-
woven containing bundled fibers (Figure 9) can
explain wicking and abrasion resistance.

Obvious tangling of fibers, as shown in Figure 10
can explain tensile properties.

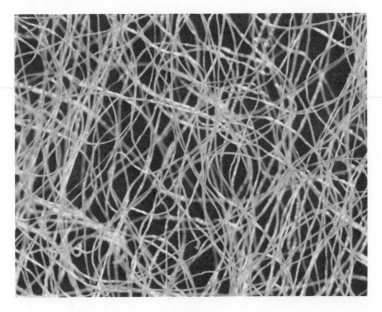

Figure 8

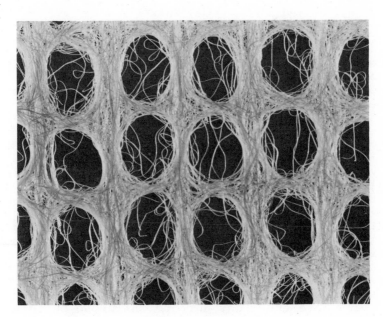

Figure 9

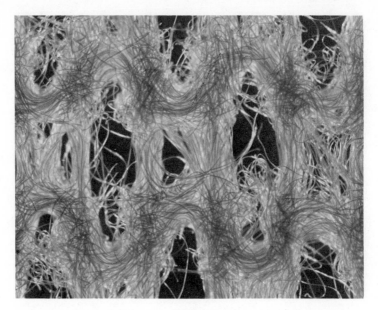

Figure 10

Where the non-woven is bonded by chemical
adhesives, the distribution and local concentration
of binder helps to understand and explain performance.   Overall impregnation, Figure 11, produces
numerous bond sites.   This prevents movement of
fibers and explains why the fabric is stiff.

Intermittent bonding produces fabrics with
many hinge points, hence a more flexible hand.   For
example, consider a non-woven fabric bonded with
polyvinyl acetate in a fine line pattern.   PVAc is
stained by I_2-KI to an intense brown or blue-brown
color.   Bonded areas, as well as unbonded areas,
are clearly shown in Figure 12.   This color is
fugitive, requiring that the photographic subject
be immersed in a dilute I_2-KI solution during observation or photography.

Figure 13 is an SEM micrograph of a rayon non-
woven fabric, bonded by viscose which was regenerated
to cellulose in situ after it was applied to the
fabric.   Note the light uniform cellulose deposit
in the bond area.

Figure 11

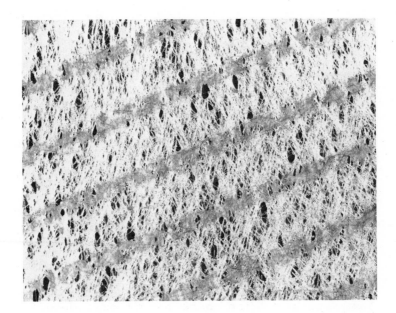

Figure 12

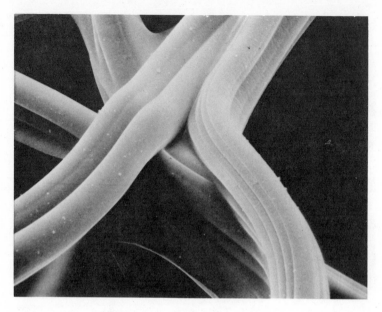

Figure 13

The fabric shown on the left side in Figure 14 is a non-woven fabric, print-bonded with an acrylic latex by the conventional method. This allows the binder to spread laterally, that is, to migrate. The fabric shown on the right was print-bonded under physico-chemical conditions which restricted binder migration producing sharper, narrower binder lines. The latter fabric has a more flexible, more textile-like hand.

Cross-sectioning further shows fiber/binder morphology. Figure 15 shows a latex-bonded fabric where capillary forces apparently controlled binder placement. The fibers evidently provided the capillary structure for the latex binder to run into. On the other hand, Figure 16, also a cross-section, shows the effect of a sudden coagulation of the latex; the fibers did not have time to fully function as capillaries because the latex was rapidly converted to a semi-solid. Under these conditions, more fibers are embedded within the solidified binder.

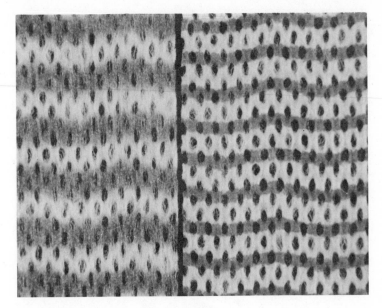

Figure 14

Figure 15

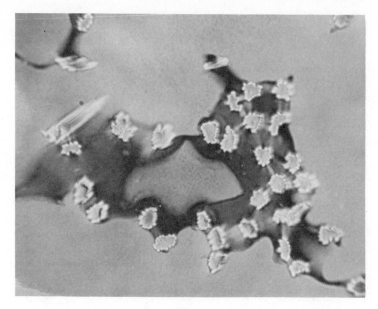

Figure 16

The microscopical approach has pitfalls. It requires a more than average skill in using equipment and interpreting the images seen. The microscopist must never forget that he has a severe sampling problem. He examines micro areas, but makes macro interpretations. As with all analytical tools, it needs to be used with care, and in conjunction with other analytical methods.

Comparison of the Properties of Ultrasonically and Mechanically Beaten Fibers

JAAKKO E. LAINE* and DAVID A. I. GORING

Pulp and Paper Research Institute of Canada and Dept. of Chemistry, McGill University, Montreal, Que., Canada

Several authors have described the ultrasonic beating of cellulosic fibers (1-4). These investigations have shown that stock consistency, sound frequency and intensity, time of treatment and extent of stirring influence the final paper properties of the fibers. These properties can differ considerably from those of mechanically beaten fibers, but the fundamental factors responsible for such differences are still not clear.

This note describes some limited observations made during experiments in which we studied changes in fiber physical properties produced by sonication (5). The results, though not comprehensive, provide some elucidation of the mechanism of ultrasonic beating.

Experimental. Bleached kraft pulp fibers from black spruce were irradiated in the previously described system (23 kHz, 10 W/cm^2) (5) for 3 h at a consistency of 0.5%. Comparative mechanical beating was done in a Valley laboratory beater according to Tappi-standard T 200 ts-66.

Both the ultrasonic and mechanical treatments were relatively mild. Thus the increase in the fines content produced by beating was considered to be negligible. Also, at such low level of beating, paper strength would be expected to be governed by interfiber bonding rather than by fiber strength.

Paper properties were determined in accordance with appropriate Tappi standards. Paper properties of the mechanically beaten sample corresponding to the same freeness and breaking length of the ultrasonically treated sample were obtained by interpolation.

Fiber saturation points were measured by the solute exclusion method as described elsewhere (5).

*Present address: Laboratory of Wood Chemistry,
                   Helsinki University of Technology, Otaniemi,
                   Finland.

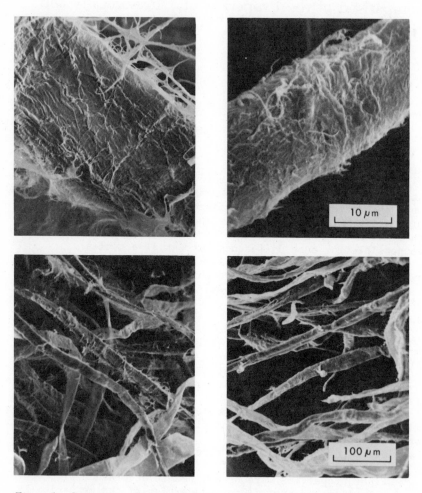

Figure 1. Comparison of scanning electron micrographs of mechanically beaten (A) and ultrasonically irradiated (B) fibers at a breaking length of 5.5 km

Results and Discussion. A comparison of the ultrasonically and mechanically beaten fibers is shown in Table I. At the same freeness ultrasonic irradiation produced higher tear, opacity and bulk than mechanical beating, but tensile and burst strengths were lower. However, comparison of the paper properties at the same tensile strength in Table II shows that the two fiber samples were similar in almost all respects. Freeness is an exception; it was considerably lower for the sonicated sample.

Table I. Comparison of mechanically and ultrasonically beaten fibers at a freeness of 528 ml.

Property	Untreated control	Type of treatment:	
		Mechanical	Ultrasonic
Tear index (mN·m^2/g)	19.6	11.7	20.8
Tappi opacity (%)	76.7	69.1	74.4
Bulk (cm^3/g)	1.75	1.51	1.60
Breaking length (km)	3.1	8.6	5.5
Burst index (kPa·m^2/g)	2.0	6.6	4.1

Table II. Comparison of mechanically and ultrasonically beaten fibers at a breaking length of 5.5 km.

Property	Untreated control	Type of treatment:	
		Mechanical	Ultrasonic
Freeness (ml)	636	610	528
Tear index (mN·m^2/g)	19.6	20.2	20.8
Tappi opacity (%)	76.7	74.0	74.4
Bulk (cm^3/g)	1.75	1.66	1.60
Burst index (kPa·m^2/g)	2.0	3.7	4.1
Fiber saturation point (cm^3/g)	0.78	0.98	1.12

Fiber flexibility is an important parameter contributing to freeness. Stone, Scallan and Abrahamson (6) have shown that the fiber saturation point as measured by the solute exclusion technique correlates with fiber flexibility. These authors found that the higher the fiber saturation point, the lower was the freeness and the higher was the flexibility of both bleached kraft and sulphite fibers. From the results in Table II, it is evident that although both types of treatment increased the fiber saturation point, the effect was larger in the case of the sonicated fibers. Therefore it is not surprising that the freeness is lower in the latter.

Scanning electron micrographs of the mechanically and ultrasonically beaten fibers are shown in Figure 1. Some surface rupture is seen in the case of the sonicated fibers confirming the results reported previously (5). However, the mechanical treatment also produced fibrillation and, for the samples shown in

Figure 1, the effect was larger in the case of the mechanically beaten fibers.

The results therefore indicate that ultrasonic irradiation increases fiber swelling and thus flexibility, thereby, providing a slow draining pulp. At equal tensile strength, mechanically beaten fibers have a higher freeness and are less swollen. Thus unusual fiber properties are produced by sonication as opposed to mechanical treatment. For conventional papermaking, a slow draining fiber is usually a disadvantage. However, there may be applications in which a swollen, flexible fiber is desirable.

Literature Cited

1. Jayme, G. and Rosenfeld, K., Das Papier, (1955), 9, (13/14), 296; (17/18), 423.
2. Kocurek, M.J., "The Mechanisms and Effects of Ultrasonic Irradiation on Cellulose Fibers", MS-thesis, State University College of Forestry, Syracuse University, Syracuse, 1967.
3. Iwasaki, T., Nakano, J., Usuda, M. and Migita, N., J. Jap. Tappi, (1967), 21, (10), 557; A.B.I.P.C., 38, 7608.
4. Laine, J.E., MacLeod, J.M., Bolker, H.I. and Goring, D.A.I., Paperi Puu, (1977), 59, (4a), 235.
5. Laine, J.E. and Goring, D.A.I., Cellulose Chem. Technol., (1977), (4).
6. Stone, J.E., Scallan, A.M. and Abrahamson, B., Svensk Papperstidn., (1968), 71, (19), 687.

Prepolymer Preparation and Polymerization of Flame Retardant Chemicals in Cotton

G. M. ELGAL, R. M. PERKINS, and N. B. KNOEPFLER

Southern Regional Research Center, Agricultural Research Service,
U.S. Dept. of Agriculture, New Orleans, La. 70179

Over several decades, research by both government and industry has resulted in a number of flame retardant finishes for textiles. Much of the current research is directed towards improved processing, with emphasis on minimizing pollution and consumption of energy. Flame retardant finishes that are available for meeting government flammability regulations for cotton textiles are based on only three basic chemicals—condensate of bis(betachloroethyl) vinyl phosphonate and alkyl phosphonate, N-methylol dimethyl phosphonopropionamide, and tetrakis(hydroxymethyl)phosphonium salts (THP salts). Several manufacturers offer THP salts and condensation products of urea and THP salts under different trade names. Recently, most of the cotton flannelette for children's sleepwear has been finished with a THP salt or condensation product by an ammonia gas cure (1). The ammonia cure process is preferred because the finished fabric has high strength retention and a soft hand. Release of unpleasant odors and design of the gaseous reactor for production scale are problems that have been encountered with this process.

The object of this study was to reduce or eliminate processing problems through modifications in treating formulations and processing operations, while maintaining product performance characteristics. Exploratory data are presented in this report.

Comparison of Processes

A schematic of the conventional ammonia gas cure process is shown in the upper half of Figure 1. The fabric is impregnated with the THP compound; the excess liquid is squeezed out by pressure rollers; and the fabric is dried to about 10–20% residual moisture at low temperature (100°C). The impregnated fabric then enters an ammonia reactor, where the ammonia gas reacts with the THP compound and forms an insoluble polymer. An oxidation step follows, which removes unpleasant odors from the finished fabric

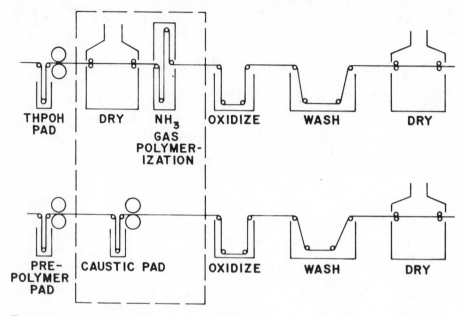

Figure 1. *Comparison of conventional THPOH–NH₃ flame retardant process and modified process*

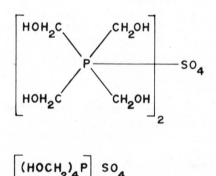

Figure 2. *Chemical structure of tetrakis(hydroxymethyl)phosphonium sulfate (THPS) salt*

and increases the durability of the finish. The fabric is then subjected to process washing and drying.

A schematic of the modified process projected by this reaseach is shown in the lower half of Figure 1. The fabric is impregnated with a THP prepolymer [2] and the excess liquid is squeezed out by pressure rollers. The impregnated fabric is passed through a strong base which causes the prepolymer to polymerize (or precipitate). This step is followed by squeeze rolling, oxidizing, washing, and drying operations. The major differences between the two processes are:

1. The use of a prepolymer rather than a THP salt in the treating formulation,
2. elimination of a drying operation after impregnation, and
3. the use of a pH change rather than gaseous ammoniation for the formation of the flame retardant finish.

THPS Chemical Structure

It was stated in the introduction that the majority of flame retardant finishes for cotton are based on THP salts. Currently, the most widely used salt is tetrakis(hydroxymethyl)phosphonium sulfate (THPS). The chemical structure of THPS is shown in Figure 2; four methylol groups are associated with one phosphorus atom. Commercially, THPS is sold as a 70-80% aqueous solution and is ionized into THP and sulfate ions. Other salts that have been produced and studied include the oxalate, a mixture of phosphate-acetate, and the chloride (1).

Prepolymer Preparation

In the conventional procedure, the THP salt is neutralized with sodium hydroxide (NaOH) to approximately pH 7, mixed into a treating formulation, and padded onto the cotton fabric (3). For the modified procedure investigated in this study, phosphoric acid is added to the THPS solution to further acidify it to approximately pH 1. Gaseous ammonia, ammonium salts, or a combination of the two are then added to the acidified THPS. Ammonia reacts with the THP salt; a precipitate is formed which immediately dissolves into the solution forming what we refer to as the prepolymer, but which could be a dissolved polymer, as shown in equation 1.

$$(HOCH_2)_4P^+ + NH_3 \xrightarrow{H^+} (-HNCH_2)P^+ + H_2O \qquad (1)$$
$$\underset{(CH_2OH)_{3-x}}{|}$$

Re-precipitation is effected by the addition of base. The prepolymer reported here somewhat resembles the adduct polymers reported by Daigle, Pepperman, and Vail (4,5).

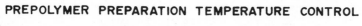

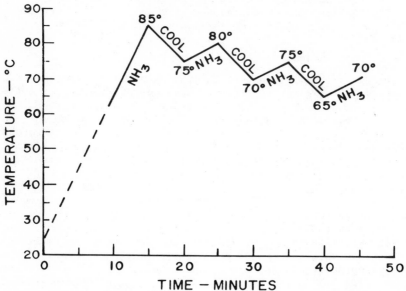

Figure 3.   *Temperature variations during prepolymer preparation*

Numerous variations have been tried for the prepolymer preparation, and each has been successful. Three representative preparations are shown in Table 1. The basic ingredients are THPS solution, phosphoric acid (H_3PO_4), and ammonia (NH_3).

TABLE I

PREPOLYMER PREPARATION

Materials	PP(1)	PP(2)	PP(6)
THPS(75%)	100g	100g	100g
H_3PO_4 (85%)	24.4g	24.4g	-
$(NH_4)_2HPO_4$	-	14.8g	14.8g
$(NH_4)_2SO_4$	-	-	14.8g
Method	Bubble NH_3 gas to saturation while stirring and cooling		Heat at 70°C for 2 min.

Ammonia gas can be used alone or in combination with an ammonium salt such as dibasic ammonium phosphate; alternatively, ammonium sulfate and dibasic ammonium phosphate can be used and the phosphoric acid omitted. When the ammonium salts are used, the mixture must be heated and kept near the boiling point for 2 minutes. In contrast, when ammonia gas is used, the overall reaction is exothermic, and the solution must be cooled when it reaches the boiling point. Formation of byproducts may be the primary cause of the temperature increase. Overheating during prepolymer preparation can reduce the reactivity of the prepolymer [6].

A typical preparation, using the ammonia gas technique, is as follows:
1. Add about 25 g of H_3PO_4 to 100 g THPS solution (about 75% solids)
2. heat the acidified THPS solution to about 60°C,
3. bubble NH_3 gas into the solution until boiling (approximately 85°C),
4. quickly cool in ice bath to 75°C,
5. return to bubbling NH_3 gas again, etc., as shown in figure 3.

Three cycles are normally sufficient to produce the desired prepolymer solution. Finally, the prepolymer should be cooled to ambient as quickly as possible to minimize deterioration.

A quick test has been used to determine completeness of prepolymer formation. Five milliliters of prepolymer is placed in a test tube; 3 to 5 ml of caustic soda (25%) is added, and the mixture is stirred with a spatula. A good prepolymer will flocculate and form a solid mass of white polymer, which can be rinsed with water without dissolution. If, however, the solution turns milky and foams upon addition of base, the prepolymer preparation is incomplete. One explanation is that some molecules remain containing $(CH_2OH)_2$-P- groups, which react with excess base to form hydrogen gas.[2] This same test can be used to determine the reactivity of a prepolymer after ageing, or after preparation with ammonium salts if overheating is suspected.

Although test tube experiments indicated immediate reaction of the prepolymer upon addition of base, some further insight into the reaction mechanisms was sought. A 5 ml beaker was half filled with 50% NaOH (Figure 4a) and prepolymer solution was carefully poured on top of the caustic producing a distinct liquid interface. A thin layer of material (about 0.5 mm thick) formed immediately; within 1 minute there was evidence of diffusion with the formation of "stalagmites" and "stalactites" below and above the interface (Figure 4b). Within two minutes the prepolymer had diffused throughout the caustic and had polymerized (or precipitated) (Figure 4c); within 5 minutes the reaction appeared complete and uniformly distributed (Figure 4d). This experiment was repeated at 30°C and 50°C and no significant difference was observed in rate of reaction. From this experiment, it was concluded that a diffusion mechanism, rather than a rate of reaction, would be the controlling factor for reaction when the liquid process prepolymer system is used. A comparison was made between the liquid process prepolymer system and the conventional ammonia gaseous system. A test tube was partially filled with THPOH prepared by the neutralization of THPS to pH 7 and ammonia gas was injected into the test tube just above the liquid layer. Polymer formation at four progressive time intervals is shown in Figure 5. A thin skin of white polymer formed instantaneously at the interface (Figure 5b). In 1 minute the polymer was 1 mm thick (Figure 5C), and in 2 minutes about 2 mm thick (Figure 5d). Thereafter the polymer layer became impervious and there was no increase in polymer thickness, even when a jet of NH_3 gas was directed onto the polymer layer. From this experiment we concluded that a diffusion mechanism was again the controlling factor for the polymerization reaction.

Fabric Treatment

The next series of experiments were designed to determine whether or not the liquid process prepolymer system could be adapted to treat cotton fabric for flame retardancy. Preliminary experiments were conducted on small samples of 100% cotton flannelette. The fabric was saturated with the prepolymer, excess solution squeezed out, the impregnated fabric dried, and caustic applied to the fabric to react it. In some samples, the drying step was eliminated. Samples dried after impregnation and before application of caustic, did not have as high add-ons as undried samples.

Figure 4. Prepolymer polymerization (or precipitation): a) initial, b) after 1 min, c) after 2 min, and d) after 5 min

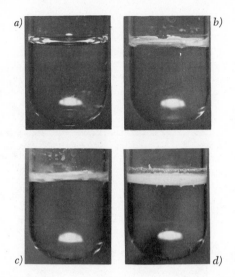

Figure 5.   Ammonia gas polymerization of THPOH:
a) initial, b) after gas addition, c) after 1 min, d) after
2 min

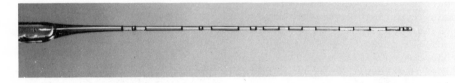

Figure 6.   Decomposition of prepolymer after oven heating for 3 min at 60°C

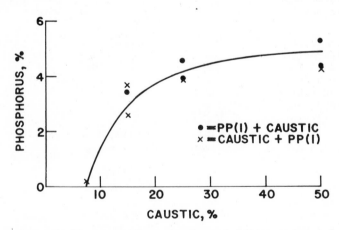

Figure 7.   Variation of flame retardancy in fabric, as measured
by phosphorus analysis, as a function of caustic concentrations

Effect of Heat on Prepolymer. An experiment was conducted
to determine the sensitivity of the prepolymer to heat. A pitot
tube was filled with prepolymer and placed in an overn at 60°C
for 3 minutes. Evidence of the decomposition of the prepolymer
and vapor formation, which tends to expel the liquid out of the
tube is seen in Figure 6. Thus, because of the sensitivity of
the prepolymer to heat, the elimination of the drying step would
be desirable.

Because the treatment with caustic is an exothermic reaction
and tends to heat the fabric during reaction, and also because
the experiment with the pitot tube verified that heating tends to
be detrimental, treatment of fabric was repeated with caustic at
0°C. In addition, the fabric was held for 1 minute in a chilled
environment before it was passed through pressure rollers. With
this chilling variation there was less contamination of the
caustic bath and the fabric had a better hand.

Effect of Caustic Concentration. Fifty percent caustic was
used in the preliminary experiments. The problems of handling
this high concentration and its cost made it desirable to
determine the minimum concentration that could be used in fabric
treatment. Data are presented in Figure 7 to show the effect of
caustic concentration on add-on, as measured by phosphorus
content in the treated fabric. The optimum appears at about 25%
caustic. In these experiments, samples were padded with prepolymer and then with caustic, or they were first padded with
caustic, dried, and then padded with prepolymer. It was theorized
that the application of the caustic first would swell the cotton
fiber and result in a higher add-on of polymer. However, it
appeared that the order of application of the prepolymer and
caustic had no significant effect on the amount of add-on. There
was a greater amount of finish wash-off into the caustic bath when
concentrations lower than 25% were used.

Effect of Prepolymer Preparation. As was mentioned earlier,
prepolymer was prepared in a variety of ways. Results are shown
in Table 2 of fabric treatments with the prepolymer prepared by 3
typical methods: ammonia gas, ammonia gas plus salt, or salt only.
The treated fabrics passed a vertical flame test (7), had add-ons
between 18.2% and 20.8%, phosphorus contents between 4.2% and
4.6%, and nitrogen contents between 1.8 and 2.7%, after one
laundry cycle. (These samples had not been given an oxidation
step.) The N/P mole ratio was 1.0 to 1.3, about the same as that
in samples treated by the conventional THPOH-NH$_3$ method (8).

Effect of Prepolymer Age. The durability of the prepolymer in
storage was investigated. Cotton flannelette was treated with
freshly prepared prepolymer and with prepolymer that had been
stored for 2 weeks and for 4 weeks. After one laundry cycle, the
percent add-on, as seen in Table 3, varied from 13.2% to 18.2%,

TABLE   2

INFLUENCE OF METHOD OF PREPOLYMER
PREPARATION ON FABRIC PROPERTIES

Method of Preparation	Add-on %	P %	N %	N/P Mole Ratio	VFT Inches
NH$_3$	20.1	4.6	2.7	1.3	1.8
NH$_3$ + Salt	18.2	4.2	1.8	1.0	2.4
Salt	20.8	4.2	2.2	1.2	2.4

Flannelette treated with prepolymer, then caustic.
Analysis made on non-oxidized samples after one laundering.

TABLE   3

INFLUENCE OF AGE OF PREPOLYMER ON FABRIC PROPERTIES

Age of Prepolymer	Add-on %	P %	N %	N/P Mole Ratio	VFT Inches
Fresh	18.2	4.2	1.8	1.0	2.4
2 weeks	17.2	3.3	1.6	1.1	BL
4 weeks	13.2	3.0	1.7	1.2	2.5

Flannelette treated with prepolymer (2) then with caustic.
Analyses made on non-oxidized samples after one laundering.

phosphorus contents were from 3.0% to 4.2%, and nitrogen contents
were from 1.6% to 1.8%. The N/P mole ratio remained constant at
1.0 to 1.2. Screening vertical flame test results indicate some
loss in the effectiveness of the prepolymer at 2 weeks; however,
samples treated with the same prepolymer after 4 weeks storage
had no failures. These variations may have been due to non-
uniform treatment of fabric; because similar data were accumulated
for twill samples and all of them passed the vertical flame test.
These results are also for non-oxidized samples.

When excess caustic was added to a 4 g sample of freshly
prepared prepolymer, about 4-6 1/2g of white polymer (or
precipitate) was formed. This quantity was gradually reduced as
the prepolymer aged and lost its reactivity and appeared dependent

upon method of preparation. For better stability, the prepolymer should be stored in opaque containers, away from light, at 10°C, and be free of contamination. These studies are continuing.

Summary

Preliminary studies were undertaken to minimize problems encountered with the conventional THPOH-NH$_3$ process when formulations were altered and processing was modified. A prepolymer was prepared by acidifying the THP salt (THPS) with phosphoric acid to pH 1, then adding ammonia to form a soluble phosphorus-nitrogen prepolymer. After the prepolymer was applied to cotton fabric, it was made insoluble by the addition of a strong base (25% NaOH). There was no drying step between application of prepolymer and caustic.

The P-N prepolymer was prepared in various ways, applied to cotton fabric, and polymerized (or precipitated) by the addition of caustic. The N/P mole ratio was approximately 1 in the treated fabric. Parameters examined were effect of heat on prepolymer, and effect of caustic concentration, prepolymer preparation, and prepolymer age on polymerization (or precipitation).

Abstract

The commercially available flame retardant chemical THPS (tetrakis(hydroxymethyl)phosphonium sulfate) was processed into a prepolymer by first acidifying with an acid or acid salt and then ammoniating under controlled conditions. After the phosphorus-nitrogen prepolymer was applied to the cotton fabric, polymerization (or precipitation) occurred with the addition of base such as sodium hydroxide. The process described is intended as a substitute for the currently used process, in which the THPS is alkalized prior to application to cotton cloth and polymerized by exposing the impregnated fabric to gaseous ammonia. The new technique eliminates the need for gaseous ammoniation, which requires specially designed ammonia reactor equipment and involves associated problems of monitoring, control, and hazardous release of ammonia into the environment. With certain alterations in the processing steps, energy could be conserved through elimination of at least one drying operation.

Literature Cited

1. LeBlanc, R. Bruce, Text. Ind. (1977), **141** (2), 29.
2. Odian, G. "Principles of Polymerization," p. 1-19, 110-124, McGraw-Hill Book Co., New York, 1970.
3. Quinn, F. J. Am. Dyest. Rep. (1974) **63** (5), 24.
4. Donaldson, D. J., Private communication.
5. Daigle, D. J., Pepperman, A. B. and Vail, S. L., U. S. Pat. No. 4,017,462, Apr. 12, 1977.

6. Daigle, D. J., Pepperman, A. B., and Vail, S. L., U. S. Pat. No. 3,961,110. June 1, 1976.
7. U. S. Fed. Supply Service, "Textile Test Methods," Fed. Spec. CCC-T191b, Method 5902, U. S. Gov. Printing Office (1951).
8. Reeves, W. A., Perkins, R. M., Piccolo, B., and Drake, G. L., Jr. Text. Res. J. (1970), 40 (3), 223.

INDEX

67783